Your SAT MATH Prep Companion

Yancy Dennis, Ph.D. & Martine Dennis

About the authors

Martine Dennis is a veteran teacher with over 20 years' experience, having taught in some of the top ranked states in the nation for education. Dr. Dennis is a Massachusetts Institute of Technology and University of Virginia alumnus with a PhD in Chemical Engineering and a NASA Fellow who achieved a perfect score on the SAT. Together they bring a wealth of knowledge and expertise both in the subject matter and pedagogy of teaching.

We strongly believe one of the primary reasons students struggle ON THE SAT is because they have been programmed to use the same APPROACHES taught in the classroom.

Our goal is to reprogram the students' thinking by providing them with multiple approaches to a problem: the traditional classroom approach versus the SAT approach. The way you approach a problem will determine how well you do on the SAT.

This book will help fill that gap by helping students get acquainted with the SAT math portion, know what to expect and illustrate best test taking strategies. This book will help you improve your speed, accuracy and boost your students' confidence by providing them with easy-to-follow, detailed step by step and time-efficient approaches to each problem. All while we try to alleviate the stress and anxiety associated with the SAT.

We recommend that you try each problem on your own and then review the various solutions for each. Also, in most cases, if a problem is taking you longer than 30 seconds, then you probably aren't using the correct or most efficient approach for solving it. Note that this book is designed to teach you the methods, strategies and techniques to score high on the actual SAT.

WHY PAY THOUSANDS OF DOLLARS when you can always take YOUR LITTLE MATH SAT COMPANION with you. This book is based on the redesigned SAT test. It is a quick and complete guide to the Math SAT. Our little SAT companion book is small and compact. It is small enough to take everywhere with you, yet complete and thorough to help you excel on the Math SAT.

Here's what you will find in this book:

- A list of all the topics covered on the Math SAT.
- A set of Math formulas and rules which are needed to do well on the SAT.
- A set of SAT type questions and strategies for each topic covered on the SAT math section
- A non-calculator test and a Calculator-allowed test that mirrors the most popular questions on the SAT
- Best Strategies for each question
- More importantly, 2 sets of approaches that compare the classroom approach vs the SAT approach.

GET TO KNOW THE SAT

The SAT test comprises 4 sections. This book focuses on the 3rd and 4th sections which are the math portions of the test. The 3rd section, the non-calculator portion, has 20 questions, 15 multiple choices and 5 grid-ins. It is 25 minutes long, which leaves you with about 75 seconds per question. The 4th section, the calculator portion of the test, has 38 questions and is 55 minutes long, roughly 87 seconds per question.

- Time management is of utmost importance.
- Monitor your time and pace yourself. You do not get extra credit for hard questions.
- All questions are worth the same number of points. Do not spend too much time on any one question.

The goal here is to use quick strategies to solve the easy questions as efficient as possible so that you have ample time to spend a minute or 2 on the harder questions. We highly recommend that you skip around and solve the easier problems first. Do not leave any questions blank. Take educated guesses if you are running out of time. You do not get penalized for wrong answers.

Strategies/Tips

- Make sure you review and study simple math calculations.
- Memorize and familiarize yourself with some math formulas
- Always read the question (Not the problem) first and scan the answer choices. It will help guide you on how to approach the problem effectively and save time.
- Skim through the problem and <u>only</u> read and underline key information to maximize your time
- Identify skills and concepts needed for that question.
- 99% of the time, you can narrow your answer choices to 2 by using key information given in the problem
- Write down the problem, show your work, write equations, draw pictures if necessary. This will help eliminate careless mistakes.
- Sometimes there/s no need to solve the problem till the end.
- Remember to substitute friendly numbers if needed.
- Review mental math tricks (multiply/divide by a power of 10, fractions, simplifying, multiply/divide numbers that end in zeros)
- Know and practice equation-solving techniques
- Eliminate fractions by multiplying every term by the lowest common denominator. It is easier and faster to work with whole numbers
- Eliminate decimals by multiplying the entire equation by a power of 10.
- Estimate
- Answer every question
- Breathe. Relax

HELPFUL FORMULAS:

- Sum of degrees in a triangle = 180
- Sum of degrees in a circle = 360
- Interior angle theorem: (n-2)180
- Sum of exterior angles in a polygon always = 360
- Perimeter of a square: 4s
- Perimeter of a rectangular: 2l + 2w or 2(l + w)
- Quadratic formula: $x = \frac{-b + \sqrt{b^2 - 4ac}}{2a}$
  - Sum of the roots: $-\frac{b}{a}, assuming\ a \neq 0$
  - Product of the roots = $\frac{c}{a}, assuming\ a \neq 0$
- Pythagorean Theorem: $a^2 + b^2 = c^2$
- Slope formula, the slope $m = \frac{y_2 - y_1}{x_2 - x_1}$
- Finding y-intercept y = mx + b, where m is the slope of the line. b is the y-intercept (the point where the line crosses the y-axis).
- Equation of a circle: $x^2 + y^2 = r^2$ if the center is at the origin. However, if the center point is (h, k), then the equation of the circle becomes $(x-h)^2 + (y-k)^2 = r^2$
- Midpoint and Distance formula
- Length of an arc L, $L = \frac{2\pi r\ x\ (measure\ of\ the\ central\ angle)}{360}$
- Area of an arc sector
- Average, Mean, Median, Range
- Trig formulas: SOH-CAH-TOA:
  - $sin = \frac{Opposite}{Hypotenuse}$
  - $cos = \frac{Adjacent}{Hypotenuse}$
  - $tan = \frac{Opposite}{Adjacent}$
- Direct Variation: $y = kx$
- Indirect Variation: $y = \frac{k}{x} or\ xy = k$
- Factored Form of a Quadratic: $y = a(x-h)^2 + k$, where the point (h, k) is the vertex.
- Exponential Functions
- Rectangular Prism: $\boldsymbol{V = lwh, Surface\ Area, SA = 2lw + 2lh + 2wh}$
- Rectangular Prism
- Density = Mass/Volume
- Rules of exponents

Let's examine what is tested on the SAT

The SAT Math section is divided into four content areas:

- **Heart of Algebra,** (HOA), consists of 19 questions which focus on the mastery of core algebraic concepts such as linear and systems of equations.
- **Problem Solving and Data Analysis,** consists of 17 questions, which focus on mathematical reasoning, linear and exponential functions, using graphs and basic statistics to analyze and summarize data
- **Passport to Advanced Math** consists of 16 questions that focus on analyzing and manipulating higher-order algebraic expressions, equations, and polynomials.
- **Additional Topics in Math** consists of 6 questions which focus on basic Geometry facts, Areas and volume calculations, circles and their equations and basic trigonometry.

Heart of Algebra questions ask students to:

1. Create, solve, or interpret a linear expression or equation in one variable that represents a context.
2. Create, solve, or interpret linear inequalities in one variable that represent a context.
3. Build a linear function that models a linear relationship between two quantities.
4. Create, solve, and interpret systems of linear inequalities in two variables.
5. Create, solve, and interpret systems of two linear equations in two variables.
6. Algebraically solve linear equations (or inequalities) in one variable.
7. Algebraically solve systems of two linear equations in two variables.
8. Interpret the variables and constants in expressions for linear functions within the context presented.
9. Understand connections between algebraic and graphical representations.

Problem Solving and Data Analysis questions ask students to:

1. Use ratios, rates, proportional relationships, and scale drawings to solve single- and multistep problems.
2. Solve single- and multistep problems involving percentages.
3. Solve single- and multistep problems involving measurement quantities, units, and unit conversion.
4. Given a scatterplot, use linear, quadratic, or exponential models to describe how the variables are related.
5. Use the relationship between two variables to investigate key features of the graph.
6. Compare linear growth with exponential growth.
7. Use two-way tables to summarize categorical data and relative frequencies and calculate conditional probability.
8. Make inferences about population parameters based on sample data.
9. Use statistics, which may involve mean, median, mode, range and/or standard deviation, to investigate measures of center of data and analyze shape, center, and spread.
10. Evaluate reports to make inferences, justify conclusions, and determine appropriateness of data collection methods.

Passport to Advanced Math questions ask students to:

1. Create a quadratic or exponential function or equation that models a context.
2. Determine the most suitable form of an expression or equation to reveal a particular trait, given a context.
3. Create equivalent expressions involving rational exponents and radicals, including simplifying, or rewriting in other forms.
4. Create an equivalent form of an algebraic expression by using structure and fluency with operations.
5. Solve quadratic equation(s) having rational coefficients.
6. Add, subtract, and multiply polynomial expressions with rational coefficients and simplify the result.
7. Solve an equation in one variable that contains radicals or contains the variable in the denominator of a fraction.
8. Solve a system of one linear equation and one quadratic equation.
9. Rewrite simple rational expressions. Students will add, subtract, multiply, or divide two rational expressions or divide two polynomial expressions and simplify the result. The expressions will have rational coefficients.
10. Interpret parts of nonlinear expressions in terms of their context.
11. Understand the relationship between zeros and factors of polynomials and use that knowledge to sketch graphs.
12. Understand a nonlinear relationship between two variables by making connections between their algebraic and graphical representations.
13. Use function notation and interpret statements using function notation.
14. Use structure to isolate or identify a quantity of interest in an expression or isolate a quantity of interest in an equation.

Remember it is almost impossible to work out every problem. However, there is a finite number of concepts covered. Try to master the concepts. It is also very helpful to identify the concept that is covered on each question then apply it.

Contents.

Section I: The Heart of Algebra

Math Topic 1: The Language and Tools of Algebra

Variables and Expressions

The difference between two numbers is 32. If the smaller number is m^4 the fourth power, then what is the greater number?

A. $m^4 - 32$
B. $m^4 + 32$
C. $32 - m^4$
D. $32m^4$

SAT Quick Solution (**Answer B**)

- Since we know the smaller is subtracted, then to find the larger number, we need to add the smaller number to 32. Thus, the final answer must have a plus sign.
- Therefore, the answer must be B.

Exponents and Order of Operations

What is the value of $\frac{m-2(y+z)}{(m-y)^2}, if\ m = -\frac{1}{2}, y = \frac{7}{2}, and\ z = \frac{9}{2}$

A. -1
B. $-\frac{33}{32}$
C. -2
D. $-\frac{65}{32}$

SAT Quick Solution (**Answer B**)

- Break down the problem and evaluate each part, then enter numbers into the expression.
- Evaluate y + z: $\frac{7}{2} + \frac{9}{2} = 8$
- Evaluate m − y: $-\frac{1}{2} - \frac{7}{2} = -\frac{8}{2} = -4$
- $\frac{-\frac{1}{2} - 2(8)}{(-4)^2} = \frac{-\frac{1}{2} - 16}{16} = -\frac{33}{2} * \frac{1}{16} = -\frac{33}{2} * \frac{1}{16} = -\frac{33}{22}$

Simplifying Algebraic Expressions

Which of the following expressions is **NOT** equivalent to $6[6y - 3(1 - y) - 5(y + 1)]$?

A. $48(y - \frac{1}{2})$
B. $24y - 48$
C. $24(y - 2)$
D. $48(\frac{1}{2}y - 1)$

SAT Quick Solution (**Answer A**)

- **TIP**: Sometimes it easier to use the answer
- Compare either the y-terms of the answer or the number parts of the answer
- Here we will go with the y-terms of the answer, but you may try the number part on your own.
    - The first expression has 48y
    - The 2nd, 3rd and 4th expressions have 24y
- Thus, the answer must by A since only one expression can be different.

Rational, Irrational, and Decimal

Which of the following expressions shows the numbers arranged in increasing order?

A. $-\sqrt{3}, -5, \frac{2}{3}, 4, \sqrt{10}$

B. $-\sqrt{3}, -5, \frac{2}{3}, \sqrt{10}, 4$

C. $\frac{2}{3}, -\sqrt{3}, \sqrt{10}, 4, -5$

D. $-5, -\sqrt{3}, \frac{2}{3}, \sqrt{10}, 4$

SAT Quick Solution (**Answer D**)

- Search for the smallest number first
- The 3 positive numbers cannot the smallest
- Next, compare -5 to $-\sqrt{3}$. Obviously -5 is smaller than $-\sqrt{3}$
- Therefore, the answer must be D.

Math Topic 2: Solving Linear Equations

Writing Equations

The quotient of *j* and *k* is 15 less than four times the sum of *j* and *k*. Which of the following equations represents the statement above?

A. $\frac{j}{k} = (4j + k) - 15$

B. $\frac{j}{k} = 15 - (4p + q)$

C. $\frac{j}{k} = 4(j + k) - 15$

D. $\frac{j}{k} = 15 - 4(j + k)$

SAT Quick Solution (**Answer C**)

- Read the question carefully.
- Focus on the elements provided in the question and relate to the answers given.
- 15 less means that we subtract 15 which eliminates **B & D**.
- From the remaining 2 answers, only C has the sum of j and k: **(j + k)**
- Therefore, the answer must be C.

Solving Equations

Solve $\frac{x}{5}+\frac{3}{4}=\frac{7}{4}x-2$

A. 2
B. $\frac{59}{31}$
C. $\frac{57}{31}$
D. $\frac{55}{31}$

SAT Quick Solution (**Answer D**)

- **TIP**: Eliminate Fractions
- **TIP**: Find Common Denominator for All Fractions in Equation
- **20 is the common denominator**
- Multiply all terms by the common denominator: $20(\frac{x}{5}+\frac{3}{4}=\frac{7}{4}x-2)$
- 4x + 15 = 35x − 40 or 35x − 40 = 4x + 15
- **TIP**: Good practice to move the largest x-term to the left-hand side of the equation
- Subtract 4x from both sides and add 40 to both sides: 31x = 55 or x = $\frac{55}{31}$

Solving Equations with variables on both sides

Solve $\frac{x-7}{7}+3=\frac{3}{14}x-2$

A. −56

B. −42

C. 42

D. 56

SAT Quick Solution (**Answer D**)

- **TIP**: Eliminate Fractions
- **TIP**: Find Common Denominator for All Fractions in Equation
- **14 is the common denominator**
- Multiply all terms by the common denominator: $14(\frac{x-7}{7}+3=\frac{3}{14}x-2)$
- 2(x−7) + 42 = 3x − 28
- 2x − 14 + 42 = 3x − 28
- 2x + 28 = 3x − 28 or 3x − 28 = 2x + 28
- **TIP**: Good practice to move the largest x-term to the left-hand side of the equation
- Subtract 2x from both sides and add 28 to both sides which yields x = 56.
- *Substituting the answers is another approach if you find solving such equations challenging.*

Equation with No Solution and Identity

For what value of m does the equation **$5(x - m) + 42 = 3x + m + 2x$** have infinite solutions?

A. 7
B. 9
C. 11
D. 12

SAT Quick Solution (**Answer A**)

- **Use the answers since you know that equation must be an identity**
- **Try 7:** $5(x - 7) + 42 = 5x + 7$
- $5x - 35 + 42 = 5x + 7$: Both the left-hand side and the right-hand side equal $5x + 7$
- Thus, 7 is the solution
- SAT Approach
    - There are infinite solutions when the left-hand side equals the right-hand side
    - Simplify
    - $5x - 5m + 42 = 5x + m$
    - $6m = 42$
    - $m = 7$

Solving for a Specific Variable

If $m = \frac{x}{x+y}$, then what is x in terms of y and m?

A. $x = \frac{my}{m+1}$

B. $x = \frac{my}{1-m}$

C. $x = \frac{my}{m-y}$

D. $x = \frac{my}{m+y}$

SAT Quick Solution (**Answer B**)

- **Pick numbers and use the answers**
- **TIP:** ALWAYS pick numbers that make calculations easy
- **Let x = 10, y = 10, then m = ½**
- Now, check answer A: $\frac{\frac{1}{2}*10}{1+.5}$. We can quickly see that this will not equal 10
- Now check answer B: $\frac{\frac{1}{2}*10}{1-.5} = \frac{5}{\frac{1}{2}} = 10.$
- Answer B is correct
- **SAT Standard Approach**
    - Cross-multiply to get m (x + y) = x or mx + my = x
    - Group like terms: $mx - x = -my$
    - Factor x: $x\ (m - 1) = -my$
    - Next, divide by $m - 1$: $x = -\frac{my}{m-1}$ $or\ x = \frac{my}{1-m}$

Math Topic 3: Functions and Linear Equations

Relations and Functions

If point (-13, b) is in Quadrant II and point (a, -7) is in Quadrant IV, in which Quadrant is point (a, b)?

A. Quadrant I
B. Quadrant II
C. Quadrant III
D. Quadrant IV

SAT Quick Solution (Answer A)

- **Approach:** Identify the sign for points a and b.
- If the first point is in Quadrant II, then b is positive
- If the second point is in Quadrant IV, then a is positive.
- Since both points are positive, then (a, b) most lie in Quadrant I
- SAT Approach
    - Draw a Picture
    - 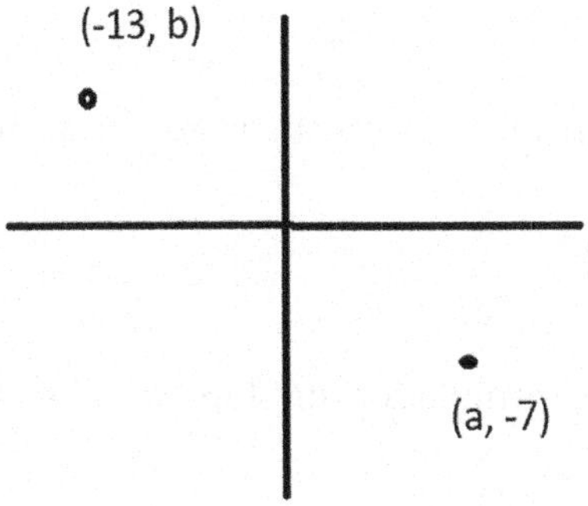

    - Examination of the graph indicates that both a and b are positive
    - Therefore, (a, b) is in the 1st quadrant.

Rate of Change and Slope

What is the slope of the line graph to the right?

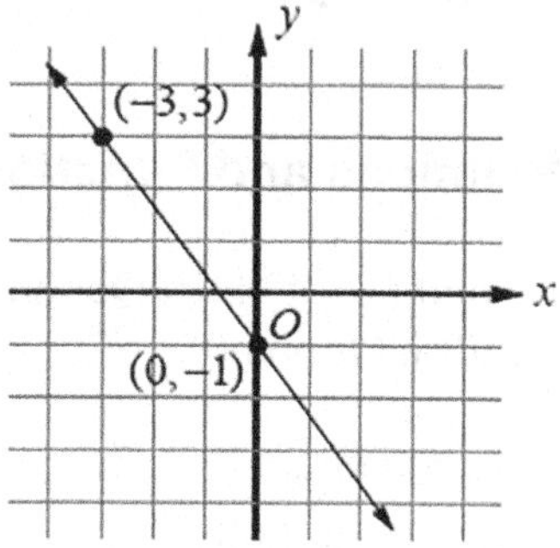

A. $-\frac{3}{4}$

B. $-\frac{4}{3}$

C. $\frac{3}{4}$

D. $\frac{4}{3}$

SAT Quick Solution (Answer B)

- **TIP:** Eliminate choices C & D, the line graphed has a negative slope
- **TIP:** Use rise over run, it's quicker than using the slope formula.

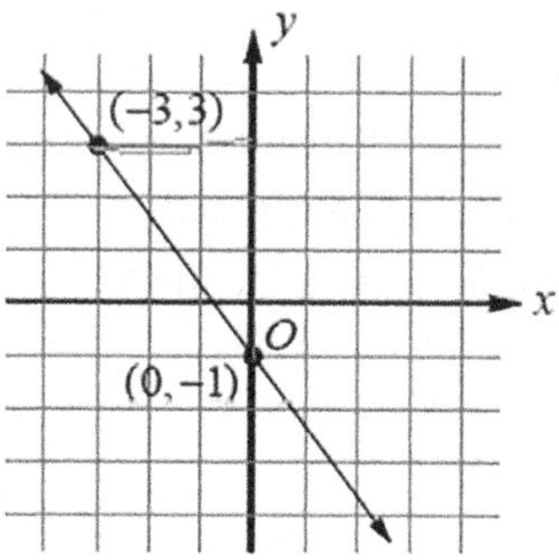

- From point 0, go up vertically 4 units, then go left 3 units to reach the point (−3, 3).
- Therefore, the slope is $-\frac{4}{3}$.
- Standard SAT Approach
    - Optionally use the formula for the slope $m = \frac{y_2 - y_1}{x_2 - x_1}$
    - $m = \frac{3-(-1)}{-3-0} = \frac{4}{-3} = -\frac{4}{3}$

Slope-Intercept Form and Point-Slope Form

Which of the following is the equation of the line that passes through point $(5, -2)$ and has slope -4?

A. $x + 4y = 18$
B. $x - 4y = 16$
C. $4x - y = 20$
D. $4x - y = 22$

SAT Quick Solution (**Answer D**)

- **Approach:** Rely on the slope-intercept form: $y = mx + b$
- **TIP:** Identify the appropriate tool to solve problems, then apply tool as needed.
- Quickly notice that lines A & B can't have a -4 slope: $4y = -x + 18$, $-4y = -x + 16$
- Should be able to check this mentally to see that dividing these lines by 4 and -4 respectively will not yield a slope of -4.
- Check C quickly, $-y = -4x + 20$, obviously the slope is 4
- Answer D is correct by elimination.
- Of course, if you can mentally rearrange the line equations, it should be apparent that only choice D has a negative slope of -4.

Parallel and Perpendicular Lines

Which of the following equations presents a line that passes through $(-6, 2)$ and is parallel to the y-axis?

A. $y = -6$

B. $y = 2$

C. $x = -6$

D. $x = 2$

SAT Quick Solution (Answer D)

- **TIP:** If the line is parallel to the y-axis, then the line must be in an equation of the form $x = b$.
- The tip eliminates choices A & B.
- Only $x = -6$ passes through the point $(-6, 2)$ because the line must coincide with the x coordinate of the point $(-6, 2)$.
- If this is unclear, then draw a quick sketch.

Solving Systems of Linear Equations

In the system of equations below, *a* and *b* are constants, and *x* and *y* are variables. If this system of equations has no solution, then what is product of *a* and *b*?

$$ax - y = 3$$
$$x - by = 12$$

A. 1

B. 2

C. −1

D. −2

SAT Quick Solution (**Answer A**)

- **TIP:** To save time, learn that the slope of a line in the form is ax + by = c is $m = -\frac{a}{b}$
- For the first line, a = a, and b = −1; therefore, the slope m = a
- For the second line, a = 1, b = −b; therefore, the slope $m = \frac{1}{b}$
- If there are no solutions, then the lines must be parallel. Consequently, the slopes must be equal such at $a = \frac{1}{b}$, therefore a and b must equal 1.
- Or you could rearrange the equations to get a * b = 1.

Absolute Value Equations

Which of the following expressions is equal to −2 for some values of x?

A. $|1 - x| + 7$
B. $|1 - x| + 5$
C. $|1 - x| + 3$
D. $|1 - x| - 3$

SAT Quick Solution (**Answer D**)

- **TIP:** Remember the absolute value is always positive
- Thus, we can eliminate A, B, and C because there is no positive number that can be added to 7, 5 or 3 to obtain −2.
- Answer D is correct by method of elimination

Math Topic 4: Linear Inequalities and Graphs

Solving Inequalities

Which of the following is the equation of inequality shown in the graph below?

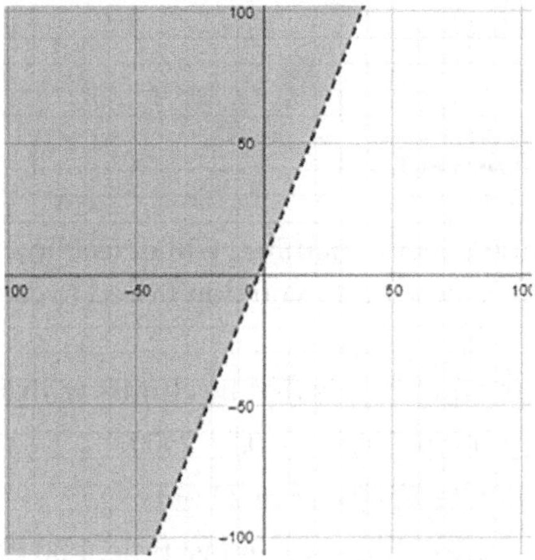

A. $5x - 2y < 10$
B. $5x - 2y > 10$
C. $2y - 5x < 10$
D. $2y - 5x > 10$

SAT Quick Solution (**Answer D**)

- The shaded area is to the left which indicates a greater than inequality
- Thus, focus on answers B & D.
- Now, use rise over run to get the slope of the graph = 50/20 = 5/2
- Rearranging B yields $-2y > -5x + 10$ or $2y < 5x - 10$ (remember to switch signs when multiplying both sides by -1)
- Rearranging D yields $2y > 5x + 10$.
- We know that the inequality is a greater because the shaded area is to the left
- Answer D is correct by method of elimination

Solving Word Problems Using Linear Models

An extreme skier is climbing up a 5900 feet high mountain. By 11:30 AM. the skier reached 900 feet and by 1:30 PM he has reached 3400 feet. If he climbs with a constant speed, by what time will he reach the top of the cliff?

A. 2:30 PM
B. 3:00 PM
C. 3:30 PM
D. 4:00 PM

SAT Quick Solution (**Answer C**)

- First, realize that the feet remaining, will detect how much time is left and that only the rate is needed so you don't need to calculate the equation of the line.
- The key clue in the question is that he **climbs with a constant speed**.
- How many feet remain: 5900 − 3400 = 2500
- Find his climbing rate: 2500 feet in 2 hours
- 2 hours remain so add 2 hours to his last time: 1:30 PM + 2 hours = 3:30 PM
- Answer C

Solving Word Problems Using Equations

In a large private school of 1240 students, two students run for president and one student receives 140 more votes than the other, how many votes did the winner receive?

A. 490
B. 640
C. 690
D. 790

SAT Quick Solution (**Answer C**)

- You could solve this set setting up an equation but let's assume that writing your own equations is not a strength.
- So, let's divide 1240 by 2 to get 620. We know the winner had to receive more than half the votes, which eliminates A.
- Let's try 640 for the winner, then the loser receives 600 votes yielding a difference of 40 votes.
- Let's try 690 for the winner, then the loser receives 550 votes yielding a difference of 140 votes.
- Answer 690.

Math Topic 5: Word Problems in Real-Life Situation

Solving Word Problems Using Systems of Equations

The star of the local basketball team makes 73 more 2-point field goals than 3-point field goals and makes a total of 216 points in the tournament.

How many 3-point goals did he score?

A. 12
B. 14
C. 16
D. 18

SAT Quick Solution (**Answer B**)

- Let's use the answers which may be easier than setting up a system of equations for some. Also, learn to understand when using the answers may be a viable approach.
- $12 \times 3 = 36$, $216 - 36 = 180$ or 90 2-point goals; $90 - 12 = 78$
- $14 \times 3 = 42$, $216 - 42 = 174$ or 87 2-point goals; $87 - 14 = 73$

Solving Word Problems Using Inequalities

The number of students in a Calculus class is four fifths the number of students in a Geology class. Both classes together have 54 students or less.

What is the greatest possible number of students in the Geology Class?

A. 28
B. 30
C. 32
D. 34

SAT Quick Solution (**Answer B**)

- Again, we can use the answer and a little reasoning. First, by the problem statement we know that there are more Geology students than Calculus students.
- So, let's try 30 Geology students which leaves 24 Calculus students. 24/30 = 4/5. Why try 30 students first? Only 30 is divisible by 5!!

Ratio and Rates

On a school trip there must be 2 chaperones for every 25 students. How many chaperones need to be on a trip if 210 students are on the trip?

SAT Quick Solution (**Answer 18 chaperones**)

- We could set up a proportion but let's examine the problem another way.
- $210/25 = 8.4$
- We must have a whole number of chaperones, so we need 9 groups in total.
- $9 \times 2 = 18$; We need 18 chaperones.

Proportions

If Lanford can drive 84 miles in 100 minutes, then how many miles can he drive in 5 hours?

SAT Quick Solution (**Answer 252 miles**)

- Change 5 hours to 300 minutes. ($5 \times 60 = 300$)
- So, the answer is $84 \times 3 = 252$ miles.

Section II: Statistics and Data Analysis

Math Topic 6: Ratios, Rates, and Proportions

Rates, Ratios and Proportions Word Problems

A 64-acre field yields 384 bushels of soybeans each year. How many more acres are needed to yield 960 bushels of soybeans each year?

A. 84
B. 90
C. 96
D. 160

SAT Quick Solution (**Answer C**)

- Bushels per acre = 384/64 = 6
- 960 / 6 = 160
- Answer: 160 − 64 = 96 acres. Read carefully. **The problem asks for how many more acres. Always read the question carefully!**

Section II: Statistics and Data Analysis

Math Topic 7: Percent

Percent of Change

If the price of a jacket increases from \$60 to \$72, then what is the price increase of the jacket?

SAT Quick Solution (**Answer 20%**)

- A quick way to do percent increase. Multiply the old price 60 by a variable, y, to get the new price, 72. 60y = 72 or y = 72/60 = 1.**20**. Keep the 2-digit number after the decimal to get the percent increase.
- So, the percent increase is 20%.
- Note, this works for percent decrease as well. For instance, if the price dropped from \$72 to \$60. 72y = 60 or y = 0.833. Thus, the price drop is 1 − 0.833 = 0.167 or 16.7%.

Percent and Equations

If Jim's monthly salary of $9,000 is 60 percent of Alison's monthly salary, what is Alison's monthly salary?

SAT Quick Solution (**Answer $15000**)

- No real trick here but use this approach
    - Replace is with an equal sign
    - Replace of with a multiplication sign
    - Which yields, $9000 = 0.60 \times y$, or $y = 9000/0.60 = \$15000$

Percent Word Problems

How many microliters of 65% acid solution must be added to 300 microliters of a 40% acid solution to make a 50% acid solution?

SAT Quick Solution (**Answer 200**)

- Set up a table to solve. **Have system to solve more complex problems.**

	Total amount x % acid = Amount of acid		
40% solution	300	40%	120
65% solution	x	65%	0.65x
New Solution	(300 + x)	50%	0.5(300 + x)

- 120 + 0.65x = 0.5(300 + x)
- 120 + 0.65x = 150 + 0.5x
- Recommend multiplying by 100 to remove decimals but not necessary
    - 12000 + 65x = 15000 + 50x
    - 15x = 3000
    - x = 200 microliters of the 65% solution must be added.

Math Topic 8: Statistics

Mean, Median, Mode, and Range

The mean of a set of n numbers is 25. If the average of the 6 greatest numbers in the set is 36 and the average of the remaining numbers is 14, what is the value of n?

A. 10
B. 12
C. 14
D. 16

SAT Quick Solution (**Answer B**)

- So, we know that the 6 greatest numbers total 6 x 36 = 216
- Approach: Use the answers
- If n is 10, then the total of the 4 remaining numbers (**4 numbers remain**) is 4 × 14 = 56. 216 + 56 = 272 which gives a set average of 272/10 = 27.2
- Try n is 12, then the total of the 6 remaining numbers is 84 (6 × 14 = 84).
    - 216 + 84 = 300
    - 300/12 = **25**
- **More standard approach requires a little more understanding**
- 6 × 36 + (n − 6) × 14 = 25n
- 11n = 132
- Answer B: n = 12

Standard Deviation

After the **Spiderman: No Way Home** movie, movie theatres at two different locations took a survey of moviegoers to determine the audience experience.

Ratings of 1000 Moviegoers from Theatre A					
Ratings	5	4	3	2	1
Frequency	280	450	110	70	90

Ratings of 1000 Moviegoers from Theatre B					
Ratings	5	4	3	2	1
Frequency	220	240	180	200	160

A. The standard deviation of the Theatre A is larger.
B. The standard deviation of the Theatre B is larger.
C. The standard deviation of the two ratings is the same.
D. The standard deviation of the two ratings cannot be determined with the data provided.

SAT Quick Solution (**Answer B**)

- This solution boils down to understanding the meaning of standard deviation. In simple terms, the more numbers are the same, the smaller the standard.
- Thus, the 4 and 5 ratings dominate the list for Theatre A. For group B, the numbers are spread out more evenly by rating, which suggests a smaller standard deviation. Thus, the standard deviation for Theatre B is larger.
- Answer B

Graphical Displays

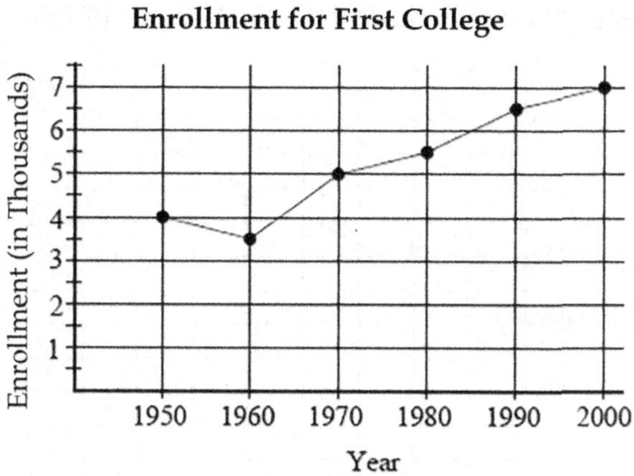

The line graph above shows the enrollment for First College from 1950 to 2000. What is the average rate of enrollment per decade between 1950 and 2000?

A. 500

B. 600

C. 750

D. 875

SAT Quick Solution (**Answer B**)

- Calculate the average rate from 1950 to 2000:
    - $\frac{7000-4000}{2000-1950} = \frac{3000}{50}$
    - But 50 years = 5 decades, so $\frac{3000}{50}$ *becomes* $\frac{3000}{5} = 600$
- Answer B
- SAT Standard Approach
    - Decade 1: −500, Decade 2: 1500, Decade 3: 500, Decade 4: 1000, Decade 5: 500
    - Sum: −500 + 1500 + 500 + 1000 + 500 = 3000
    - Average Rate of per Decade = $\frac{3000}{5}$ = 600

Scatter Plots and the Regression Lines

A. 1996

B. 1998

C. 2000

D. 2002

Using the line of best fit in the scatter plot above to estimate the best approximate year in which the average Quidditch Score for Team Wizard was approximately 4500.

SAT Quick Solution (**Answer A**)

- Look at the horizontal line on the y-axis that indicates 4500 points scored
- Visualize a vertical line down to the x-axis at the intersection.
- The vertical line intersects the vertical line at approximately 1996.
- Answer A

Populations, Samples, and Random Selection

A bag contains 21 balls numbered 1 to 21. If a ball is selected at random and then replaced, what the probability that the first ball selected is a prime number and the second ball selected is a multiple of 3?

SAT Quick Solution (**Answer:** $\frac{8}{63}$)

- Identify the prime numbers between 1 and 21: 2, 3, 5, 7, 11, 13, 17, 19
- Identify the multiples of 3 between 1 and 21: 3, 6, 9, 12, 15, 18, 21.
- Probability of a prime number: 8/21
- Probability of a multiple of 3: 7/21 = 1/3
- To get the probability of events A & B, multiply the 2 probabilities:
    - 8/21 × 1/3
    - 8/63

Math Topic 9: Categorical Data and Probability

Permutations and Combinations

Donovan has seven different pairs of sneakers in his closet. If he keeps his Air Jordan's in the middle, how many ways can he arrange those 7 pairs of sneakers?

A. 120

B. 600

C. 720

D. 5040

SAT Quick Solution (**Answer C**)

- Since the Air Jordan's remain in the middle, then we are only rearranging 6 pairs of sneakers in effect.
- $6! = 6 \times 5 \times 4 \times 3 \times 2 \times 1 = 720$ possible arrangements

Probability

In a box of 24 airpods 3 are defective. If you choose two airpods without replacement, what is the probability that both airpods are defective?

A. $\frac{1}{368}$

B. $\frac{1}{192}$

C. $\frac{1}{108}$

D. $\frac{1}{92}$

SAT Quick Solution (**Answer** $\frac{1}{92}$)

- Since we are looking at the probability that both pairs are defective, we can assume that we selected a defective airpod for each selection.
- The probability of a defective airpod on the first selection is 3/24 or 1/8.
- The probability of a defective airpod on the second selection is 2/23.
- Thus, the probability that both are defective is $\frac{1}{8}\, x \frac{2}{23} = \frac{1}{92}$

Categorical Data and Conditional Probabilities

Based on the table below, what is the probability that a randomly chosen student is a math major given that the student is a male?

	History	Math	French	Total
Male	300	600	50	950
Female	450	800	100	1350
Total	750	1400	150	2300

SAT Quick Solution (**Answer** $\frac{12}{19}$)

- First, find the total number of male students: 950. This satisfies the condition that the student is male.
- Next, find the total number of male math students: 600
- Probability = $\frac{600}{950} = \frac{12}{19}$

Section III: Advanced Math

Math Topic 10: Law of Exponents and Polynomials

Laws of Exponents and Scientific Notation

If $2^x = 7$, then what is the value of $2^x + 2^{2x} + 2^{3x}$?

SAT Quick Solution (**Answer 399**)

- Need to Know: $(a^m)^n = a^{mn}$
- Thus, $2^{2x} = (2^x)^2$ and $2^{3x} = (2^x)^3$
- Now, substitute $7 + 7^2 + 7^3 = 399$
- **Learn those rules of exponents!!**

Adding, Subtracting, Multiplying, and Dividing Polynomials

If $\frac{6x^2+6x-1}{3x+1} = 2x + \frac{4}{3} + \frac{A}{3x+1}$, Find A.

SAT Quick Solution (**Answer** $-\frac{7}{3}$)

- **Approach 1**
- Set x = 0,
- Thus −1/1 = 2(0) + 4/3 + A/1
- Or −1 = 4/3 + A
- Subtract 4/3 from both sides to get A = −7/3
- **Approach 2**
- Eliminate the denominator by multiplying entire equation by 3x + 1
- $6x^2 + 6x - 1 = 2x(3x + 1) + \frac{4}{3}(3x + 1) + A$
- Simplify right-hand side: $6x^2 + 2x + 4x + \frac{4}{3} + A = \mathbf{6x^2 + 6x + \frac{4}{3} + A}$
- The x-terms cancel each other out.
- Which reduces the equation to $\frac{4}{3} + A = -1$ *or* $A = -1 - \frac{4}{3} = -\frac{7}{3}$

FOIL Method and Special Products

What is the value of the following expression: (6 – 3x) (3x + 6)?

A. $9x^2 + 81x - 36$

B. $-9x^2 - 36$

C. $9x^2 + 36$

D. $-9x^2 + 36$

SAT Quick Solution (**Answer D**)

- Realize this is a foil problem but let's rewrite 6 – 3x to get – (3x – 6)
- **Need to know**: $(a - b)(a + b) = a^2 - b^2$
- So, now we have – (3x –6) (3x + 6) = $-(9x^2 - 36)$ = $\mathbf{-9x^2 + 36}$

Prime Factorization, GCF, and LCM

If $2 \cdot 9^9 - k \cdot 9^8 = 2 \cdot 9^8$, then what is the value of k?

SAT Quick Solution (**Answer 16**)

- Remember multiplication precedes subtraction, so write the problem applying the principle that $a^x = a \cdot a^{x-1}$
    - $2 \cdot 9^9 - k \cdot 9^8$
    - $= 2 \cdot 9^9 - (k \cdot 9^8)$
    - $= 2 \cdot 9 \cdot 9^8 - (k \cdot 9^8)$
    - $= 18 \cdot 9^8 - k \cdot 9^8$
    - Now, factor 9^8 to get
    - $(18 - k) \cdot 9^8 = 2 \times 9^8$
    - $18 - k = 2$; $\mathbf{k = 16}$

Factoring Using the Distributive Property

Which of the following is equivalent to the expression $7n^2 - 21mn - 2n + 6m$?

A. $(7n + 2)(2n - 3m)$
B. $(7n - 2)(n - 3m)$
C. $(7n - 2)(2n - 3m)$
D. $(7n - 3)(n - 2m)$

SAT Quick Solution (**Answer B**)

- Develop a systematic approach no need to foil the entire expression.
- First look at the squared first term which enables the student to eliminate choices A and C immediately. $14n^2 \neq 7n^2$
- Look at last term both B and D: 6m = 6m. No help!
- So, let's move onto the next to last term: −2n, only B gives us a −2n.
- Answer B

Math Topic 11: Quadratic Functions

Graphing Quadratic Equations

The graph of the equation y = a (x − 2) (x + 4) is a parabola with vertex (h, k). If the minimum value of y is −7, what is the value of a?

SAT Quick Solution (**Answer** $\frac{7}{9}$)

- **Need to Know**: The x-coordinate of the vertex is the average of the 2 x-intercepts
    - 2, −4
    - $\frac{2-4}{2} = -1$
- Thus, the vertex is (-1, -7)
- Next, substitute $-7 = a(-1-2)(-1+4) = a(-3)(3) = -9a$
- Thus, $a = \frac{7}{9}$

Factoring Trinomials

Factor $-x^2 + 5x + 84$.

A. $(21 - x)(x + 4)$
B. $(21 - x)(x - 4)$
C. $(12 + x)(x - 7)$
D. $(12 - x)(x + 7)$

SAT Quick Solution (**Answer: D**)

- Examine the last term: 84. It's positive. Referring to solutions provided, this eliminates B & C since the constant is −84 for these two choices.
- Next, look at the second term: 5x, examine answer A → $21x - 4x = 17x$. Eliminates A. Thus, answer must be D.
- Standard SAT Approach
  - Find a suitable factor of 84: 1, 84; 2, 42; 3, 28, 4, 21; **7, 12**; 14, 6
  - $12 - 7 = 5$ which gives the 5x as a middle term
  - $(-x + 12)(x + 7)$ or $(12 - x)(x + 7)$

Factoring Differences of Squares and Perfect Square Trinomials

If $x - y = \sqrt{7}$ *and* $x + y = \sqrt{7}$*, then what is* $x^2 - y^2 =$?

A. 7
B. $7\sqrt{7}$
C. 49
D. Cannot be determined

SAT Quick Solution (**Answer: A**)

- Quickly identify this as a difference of squares problem:
  - $(x - y)(x + y) = x^2 - y^2$
  - Substitute $\sqrt{7} \times \sqrt{7} = 7$
  - Answer: A

Solving Quadratic Equations by Completing the Square

To complete the square for the equation $x^2 - \frac{mx}{6} = 12$, what term do you need to add to both sides of the equation.

A. $\frac{mx^2}{3}$

B. $\frac{mx^2}{6}$

C. $\frac{mx}{36}$

D. $\frac{m^2x^2}{144}$

SAT Quick Solution (**Answer: D**)

- Take ½ of the x-term: $\frac{1}{2}\left(\frac{mx}{6}\right) = \frac{mx}{12}$
- Square it: $\left(\frac{mx}{12}\right)^2 = \frac{m^2x^2}{144}$
- Answer: D

Quadratic Formula and the Discriminant

What is the sum of all values of x that satisfy the equation $5x^2 - 3x - 1 = 0$?

SAT Quick Solution (**Answer: D**)

- **Need to Know**: Sum of all solutions for a quadratic equation: $-\frac{b}{a}$
- $a = 5$, $b = -3$
- $\frac{b}{a} = -\left(-\frac{3}{5}\right) = \frac{3}{5}$
- Standard SAT Approach
    - Apply the quadratic formula
    - $x = \frac{3 \pm \sqrt{(-3)^2 - 4(5)(-1)}}{2(5)}$
    - Simplify: $x = \frac{3 \pm \sqrt{29}}{10}$
    - For the sum of the solutions, we can eliminate the radical part:
        - $\frac{3}{10} + \frac{3}{10} = \frac{6}{10} = \frac{3}{5}$

Solving Systems Consisting Linear and Quadratic Equations

$$y = x^2 + x$$
$$y = ax - 1$$

In the above system of equations, a > 0. If the system of equations has exactly one real solution, what is the value of *a*?

SAT Quick Solution (**Answer: 3**)

- First, set two equations equal: $x^2 + x = 2ax - 1$
- Rearrange: $x^2 + (1 - a)\,x + 1 = 0$
- You should know that $x^2 + 2x + 1 = 0$ has only a single solution.
- Or $(x + 1)^2 = 0$. X = −1
- Thus, 1 − a = 2 or solving yields a = −1; does not meet the criteria a > 0
- Also, $x^2 - 2x + 1 = 0$ has a single solution; $(x - 1)^2$
- Thus, (1 − a) = −2
- a = 3
- Standard SAT Approach
    - First, set two equations equal: $x^2 + x = 2ax - 1$
    - Rearrange: $x^2 + (1 - a)\,x + 1 = 0$
    - **Need to Know: Exactly one solution when $b^2 - 4ac = 0$**
    - **$(1 - a)^2 - 4\,(1)\,(1) = 0$**
    - Factor $(1 - a)^2 = 4$
    - Or (1 − a) = 2 or (1 − a) = −2
    - Solving the first equation yield a value for a < 0.
    - For the second equation, a = 3.

Math Topic 12: Composition, Recursion, and Exponential Functions

Composition of Functions

If $f(x) = x^3 - 2x + 1$, and $g(x) = 1 - x$, then what is $g(f(-2)$?

SAT Quick Solution (**Answer 4**)

- Questions of this type are easy. Just understand the strategy.
- Evaluate the inside first: $f(-2) = (-2)^3 - 2(-2) + 1 = -3$
- $g(-3) = 1 - (-3) = 4$
- Answer 4

Recursive Formula

A sequence is recursively defined by $a_n = \sqrt{(a_{n-1})^4 + 3}$. If $a_o = \sqrt{3}$, then what is the value of a_2?

A. $\sqrt{12}$
B. 12
C. $\sqrt{147}$
D. 147

SAT Quick Solution (**Answer: C**)

- $a_1 = \sqrt{(\sqrt{3})^4 + 3} = \sqrt{9+3} = \sqrt{12}$
- $a_2 = \sqrt{(\sqrt{12})^4 + 3} = \sqrt{144+3} = \sqrt{147}$

Exponential Functions and Graphs

What is the horizontal asymptote of the graph of the equation $y = -2e^{-x-7} - 7$?

SAT Quick Solution (**Answer** −7)

- This is straightforward also.
- If x is very large and positive, then y approaches infinity
- If x is very large and negative, the first term, $-2e^{-x-7}$ approaches 0 because the term essentially becomes $-\frac{2}{\infty}$ which also approaches 0. Thus, the asymptote is $0 - 7$ or -7

Exponential Growth and Decay

The number of rabbits in a certain population doubles every 25 days. If the population starts with 18 rabbits, which of the following gives the total number of rabbits in the population after **d** days?

A. $18(2)\left(\frac{d}{25}\right)$

B. $18(2)(\frac{25}{d})$

C. $18(2)^{\frac{25}{d}}$

D. $18(2)^{\frac{d}{25}}$

SAT Quick Solution (**Answer D**)

- Recognize that the question requires an exponent, thus eliminate A & B.
- Recognize that the exponent cannot be $\frac{25}{d}$. Why? At 0 days, $\frac{25}{d}$ *approaches infinity* and we know that we only 18 rabbits on day 0.
- Answer D

Math Topic 13: Polynomial and Radical Functions

Polynomial Functions and Their Graphs

What real value of x satisfies the equations $x^3 - 10x^2 + 10x - 100 = 0$?

SAT Quick Solution (**Answer 10**)

- Remember that is unlikely that the SAT would ever ask you to factor a cubic equation, unless there was some strategic approach.
    - GROUP LIKE TERMS
- $x^2(x - 10) + 10(x - 10) = 0$
- Here you can quickly recognize that x equals 10 satisfies the equation.
- For COMPLETENESS the equation above becomes $(x^2 + 10)(x - 10) = 0$.
- Only the term x − 10 provides a real solution; x = 10.

Remainder Theorem and Factor Theorem

What is the remainder of polynomial $p(x) = x^5 - 100x^3 + 19x - 64 = 0$ is divided by 1+x?

SAT Quick Solution (**Answer 16**)

- You could use long or synthetic division but there's a quicker approach.
- Recognize that the question is asking what the remainder is when the equation p(x) is divided by −1.
- Or what is $p(-1)$?
    - $p(-1) = -1^5 - 100(-1)^3 + 19(-1) - 64 = -1 + 100 - 19 - 64 = 16$

Radical Expressions

Which of the following is equal to $m^{5/3}$?

A. $m\sqrt{m}$

B. $m\sqrt{m^{\frac{1}{3}}}$

C. $m\sqrt[\frac{1}{3}]{m}$

D. $m\sqrt[3]{m^5}$

SAT Quick Solution (**Answer: D**)

- Remember $x\sqrt{y^m} = x^{m/n}$. Thus, only D is equal to $m^{5/3}$
- Remember that you add exponents: $1 + \frac{2}{3} = \frac{5}{3}$
- Since every answer has an m, we only need to compare the radical parts.
    - $\sqrt{m} = m^{\frac{1}{2}} \neq m^{\frac{2}{3}}$, Eliminate A
    - $\sqrt{m^{\frac{1}{3}}} = m^{\frac{1}{6}} \neq m^{\frac{2}{3}}$, Eliminate B
    - $\sqrt[\frac{1}{3}]{m} = m^3 \neq m^{\frac{2}{3}}$, Eliminate C
    - Thus, the answer must be D.

Solving Radical Equations

What is the solution set of the equation $\sqrt{5x - 12} = 3\sqrt{2}$?

A. 2
B. 4
C. 6
D. 8

SAT Quick Solution (**Answer C**)

- **Use Answers**
  - Eliminate A: **5(2) − 12 = -2**, can't take the square root of a negative number
  - Try 4: **5(4) − 12 = 8**, the square root of 8 is not equal to $3\sqrt{2}$
  - Try 6: 5(**6**) **− 12 = 18**, the $\sqrt{18} = \sqrt{9(2)} = 3\sqrt{2}$; Answer C
- **Alternative Approach**
  - Or Square both sides: 5x − 12 = $3^2(\sqrt{2})^2 = 9 * 2 = 18$
  - Solving 5x − 12 = 18 yields 5x = 30 or x = 6

Complex Numbers

What is the value of $\sqrt{-4}\sqrt{-9}$?

SAT Quick Solution (**Answer 6**)

- Do you really need to apply the rules of complex numbers?
  - No
  - Remember $\sqrt{x}\,\sqrt{y} = \sqrt{x \cdot y}$
- Thus, $\sqrt{-4}\sqrt{-9} = \sqrt{-4 \cdot -9} = \sqrt{36} = 6$
- **Tip 1**: REVIEW COMPLEX NUMBERS, PROBLEMS ARE TYPICALLY EASY LEARN THE RULES
- **Tip 2**: SOMETIMES YOU CAN AVOID COMPLEX NUMBERS AS IN THE ABOVE EXAMPLE

Math Topic 14: Rational Expressions

Rational Expressions

Given that x ≠ 1, then the expression $\frac{1-\frac{1}{x+1}}{1+\left(\frac{1}{x^2-1}\right)}$ is equivalent to?

A. $\frac{x+1}{x}$

B. $\frac{x-1}{x}$

C. $\frac{x+1}{x^2}$

D. $\frac{x-1}{x^2}$

SAT Quick Solution (**Answer B**)

- If you have difficulty simplifying complex expressions, then select a value for x and plug that value into the choices until the two are equal.
- Let x = 2, then $\left(1 - \frac{1}{2+1}\right) = 1 - \frac{1}{3} = \frac{2}{3}$ for the numerator
- For the denominator: $1 + \left(\frac{1}{4-1}\right) = 1 + \frac{1}{3} = \frac{4}{3}$
- So, when x = 2, the expression is $\frac{\frac{2}{3}}{\frac{4}{3}} = \frac{2}{\cancel{3}} x \frac{\cancel{3}}{4} = \frac{2}{4} = \frac{1}{2}$
- Plug 2 into choice A: (2 + 1)/2 = 3/2; wrong answer
- Plug 2 into choice B: (2 − 1)/2 = ½; right answer

Solving Rational Equations

For what value of x is the expression $h(x) = \frac{1}{(x+5)^2 - 24(x+5) + 144}$ undefined?

SAT Quick Solution (**Answer 7**)

- You could expand the denominator, set it equal to 0 and solve for x.
- However, there is a more practical approach requiring much less Algebra.
- Let y = x + 5 which avoids the exercise of expanding and simplifying the denominator
- Therefore, substitute y for x +5 and set the expression equal to 0
    - $y^2 - 24y + 144 = 0$
    - Factorization yields $(y - 12)(y - 12) = 0$
    - $y - 12 = 0$ or $y = 12$
    - $y = x + 5$; $12 = x + 5$; $x = 7$

Direct, Inverse, and Joint Variations

If y varies inversely as x, and y = 8 when x = 25, what is the value of y when x = 100?

A. 1

B. 2

C. 3

D. 4

SAT Quick Solution (**Answer B**)

- Direct variation: y = kx
- Indirect Variation: $y = \frac{k}{x}$ *or* $xy = k$
- We are given x and y: (8)(25) = k or k = 200
- So, the general equation becomes xy = 200
- We are given that x = 100, so y = 2

Solving Word Problems using Rational Expressions

Three machines, A, B and C, working together at their respective constant rates, can finish a job in 9 hours. Machines A and B working together can finish the same job in 12 hours. How many hours will it take machine C working alone to finish the job?

A. 20

B. 24

C. 28

D. 36

SAT Quick Solution (**Answer D**)

- You know that machine C is slower than A & B. Why? If we assume A & B work at the same rate, it takes each machine 24 hours to finish the job.
- So, in 9 hours, A & B can finish approximately ¾ (9/12) of the job which leaves ¼ of the job for C.
- Thus, $\frac{9}{\frac{1}{4}} = 36\ hours$. Answer D
- **SAT Approach**
- Or solving the problem directly as below:
    - $\left(\frac{1}{a}+\frac{1}{b}\right)=\frac{1}{12}\ and\ \left(\frac{1}{a}+\frac{1}{b}+\frac{1}{c}\right)=\frac{1}{9}$
    - $\left(\frac{1}{12}+\frac{1}{c}\right)=\frac{1}{9}\ or\ \frac{1}{c}=\frac{1}{9}-\frac{1}{12}=\frac{1}{36}$
    - Thus, c = 36

Math Topic 15: Trigonometric Functions

Trigonometric Ratios of Acute Angles

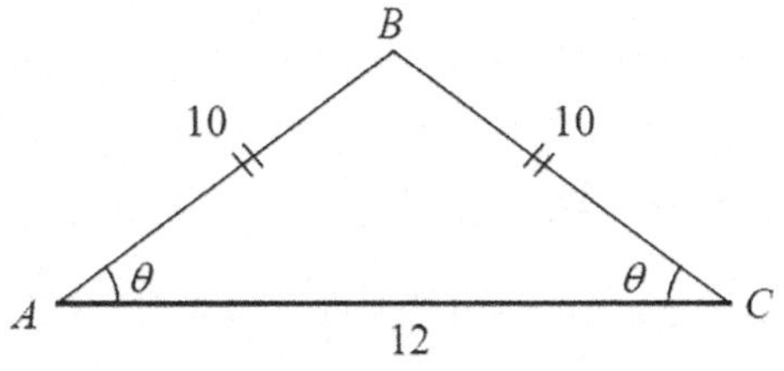

In the figure above, what is the tan θ?

A. ¾

B. $\frac{4}{3}$

C. $\frac{5}{3}$

D. $\frac{5}{4}$

SAT Quick Solution (**Answer B**)

- Split the triangle and half and you can calculate the length of the shorter leg by applying the 6-8-10 Pythagorean triple.
- Tangent θ is opposite over adjacent: $\frac{8}{6} = \frac{4}{3}$
- Answer B

The Radian Measure of an Angle

Which of the following is equal to cos ($\pi/12$)?

A. sin ($\pi/12$)
B. sin ($\pi/6$)
C. sin ($\pi/4$)
D. sin ($5\pi/12$)

SAT Quick Solution (**Answer D**)

- Need to know that cos (θ) = sin (90 − θ)
- 90 degrees is equivalent to $\frac{\pi}{2}$
- $\frac{\pi}{2} - \frac{\pi}{12} = \frac{6\pi}{12} - \frac{\pi}{12} = \frac{5\pi}{12}$
- *Answer D*

Trigonometric Functions and the Unit Circle

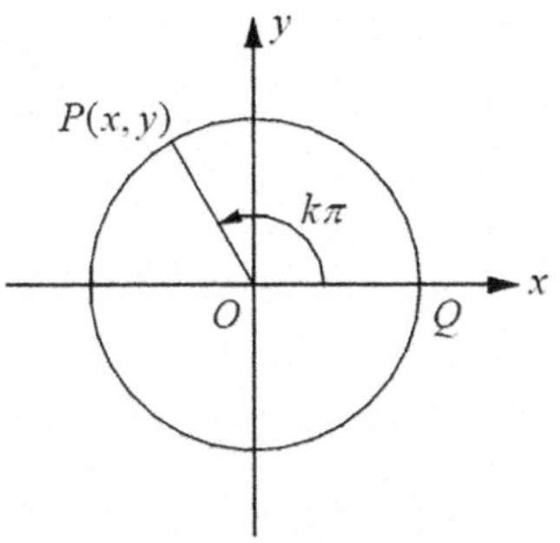

If the sin (kπ) = a, then what is the sin (kπ + π)?

A. ½ a

B. −½ a

C. a

D. −a

SAT Quick Solution (**Answer D**)

- 2nd approach requires a little understanding of trigonometry but is even easier
- kπ terminates in the second quadrant. If we add 180 or π degrees, it terminates in the 4th quadrant.
- The sin is positive in the 2nd quadrant and negative in the 4th quadrant. Thus, the sin (kπ + π) = −a
- **SAT Standard Approach**
    - If you take the time to remember a few extra formulas, then problems of this type are rather straightforward, assuming that you have a basic knowledge of trigonometry
        - sin (a + b) = sin a × cos b + cos a × sin b
        - cos (a + b) = cos a × cos b + sin a × sin b
    - Thus, a = kπ and b = π
    - Apply the first formula. We know that sin π = 0, which eliminates the 2nd term.
    - Given sin a = sin(kx) = a
    - cos b = cos π = −1
    - Therefore, sin (kπ + π) = a · (−1) = −a

Section IV: Geometry

Math Topic 16: Lines and Angles

Lines, Segments and Rays

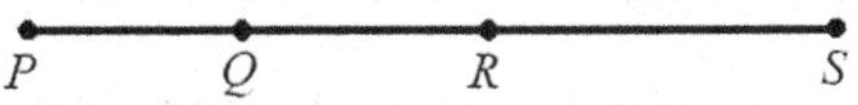

The Figure is not drawn to scale.

On the segment PS above, PR = 24, QS = 32, and QR = $\frac{PS}{3}$. What is the length of PS?

A. 38
B. 40
C. 41
D. 42

SAT Quick Solution (**Answer: D**)

- Since PS = 3 QR, it's safe to assume that PS is a multiple of 3.
- The only answer that is a multiple of 3 is D, 42.
- **SAT Approach**
    - $PR = PQ + QR$
    - $QR = \frac{1}{3}PS$
    - $24 = PQ + \frac{1}{3}PS$
    - $Solve\ for\ PQ\colon PQ = 24 - \frac{1}{3}PS$
    - $QS = QR + RS, QS = 32, QR = \frac{1}{3}PS$
        - $32 = \frac{1}{3}PS + RS$
    - $RS = 32 - \frac{1}{3}PS$
    - $PS = PQ + QR + RS$
    - $PS = \left(24 - \frac{1}{3}PS\right) + \frac{1}{3}PS + (32 - \frac{1}{3}PS)$
    - $PS = 56 - \frac{1}{3}PS$
    - $\frac{4}{3}PS = 56$
    - $PS = 56\ x\ ¾ = 42$
    - Wow, a difficult problem by the standard approach!!

Angles

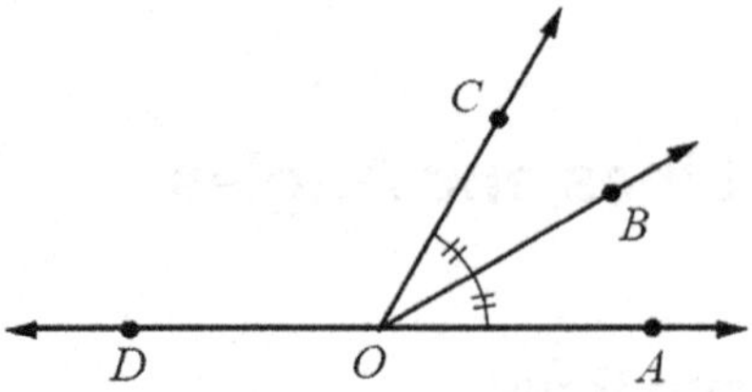

The figure is not drawn to scale

In the figure above, the ray OB bisects angle COA. If the measure of angle of DOB = 22x + 12 and the measure of angle COA = 16x − 24, then what is the measure of angle DOC?

A. 92
B. 96
C. 102
D. 108

SAT Quick Solution (**Answer D**)

- Apply straight angle theorem: $m\angle DOA = m\angle DOB + m\angle BOA$
- $m\angle BOA = ½\, m\angle COA = ½\,(16x - 24)$
- 180 = 22x + 12 + ½ **(16x − 24)**
- 180 = 22x + 12 + 8x − 12 = 30x; 30x = 180
- x = 6
- Therefore DOC = DOA - COA
- 180 − (16x − 24) = 180 − (16×6 − 24) = 180 − (96 − 24) = 180 − 72 = **108**
- **Answer: D**

Parallel and Perpendicular Lines

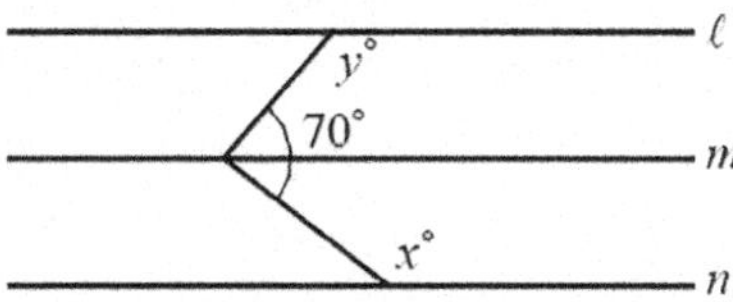

The figure is not drawn to scale

In the figure above, lines *l, m* and *n* are parallel. What is the measure y°?

A. 90
B. 120
C. 135
D. 145

SAT Quick Solution (**Answer D**)

- Even though a figure is not drawn to scale, it is difficult to draw a figure that is completely off scale. Thus, it is reasonable to assume that angle y is greater than 90°. Eliminate A.
- Line *m* bisects the 70° angle. Therefore, the angle on either side of the line *m* is 35°.
- The 35° and angle y are same side interior angles and therefore must add to 180°.
- Thus, the measure of angle y = 180 − 35 = 145°.

Parallel and Perpendicular Lines

The figure is not drawn to scale.

Math Topic 17: Triangles

Angles of a Triangle

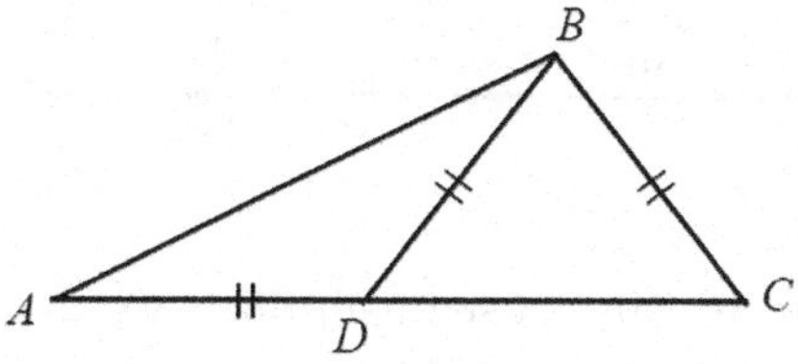

In the figure above, AD = BD = BC. If the measure of angle A is 28, then what is the measure of angle DBC?

A. 66

B. 68

C. 70

D. 72

SAT Quick Solution (**Answer B**)

- If the measure of angle of A is 28, then the measure of angle B must also be 28.
- Therefore, the measure of angle ADB = 180 − 28 − 28 = 124
- Angle ADC is a straight angle, thus ADB + BDC = 180
- 124 + BDC = 180 or BDC = 56
- Thus, the measure of angle DBC is 56 + 56 + DBC = 180
- DBC = 180 − 56 − 56 = 68

Pythagorean Theorem and Special Right Triangles

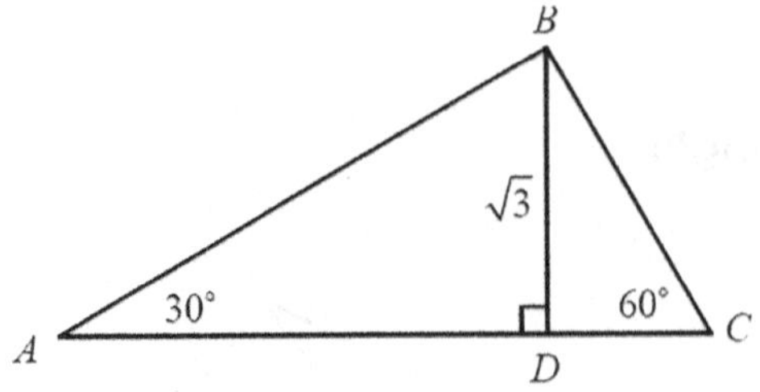

In the figure above, the measure of segment BD is $\sqrt{3}$. What is the perimeter of triangle ABC?

A. $2\sqrt{2} + 6$

B. $2\sqrt{3} + 6$

C. $2\sqrt{6} + 6$

D. $3\sqrt{2} + 6$

SAT Quick Solution (**Answer B**)

- Apply rules of special right triangle: 30-60-90
  - The length of the long leg opposite the 30 degrees is $\sqrt{3} \times$ length of the short leg
  - The length of the hypotenuse is 2x the length of the short leg.
- If you know that the $\sqrt{2}$ is associated with 45-45-90 triangles, then eliminate A & D.
- $\sqrt{6}$ is unlikely to part of the answer because we are given an answer with the $\sqrt{3}$ and based on the rules of a 30-60-90 there is really no way that $\sqrt{6}$ is part of the answer. T**hus, the answer must be B.**
- However, we will solve for students who don't want to work their way to the answer using mathematical reasoning.
- Thus, the length AD is $\sqrt{3} \times \sqrt{3} = 3$
- The length of AB is $2 \times \sqrt{3} = 2\sqrt{3}$
- The length of DC is $\frac{\sqrt{3}}{\sqrt{3}} = 1$ because BD is the longer leg of triangle BDC
- The length of the hypotenuse of triangle BDC $= 2 \times 1 = 2$
- Therefore, the perimeter $= 3 + 2\sqrt{3} + 1 + 2 = 6 + 2\sqrt{3}$

Similar Triangles and Proportional Parts

Find x.

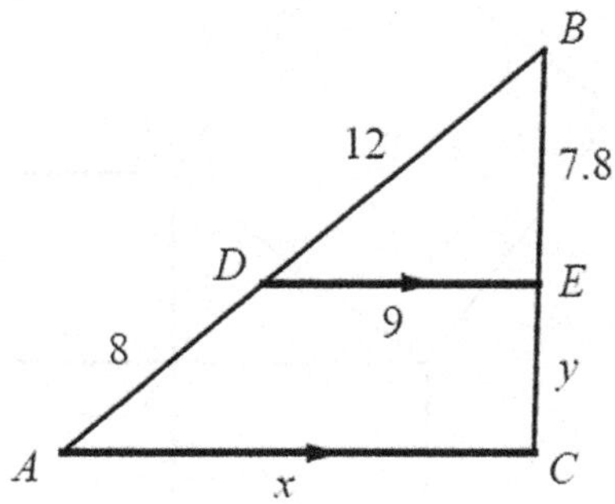

A. 12
B. 14
C. 15
D. 18

SAT Quick Solution (**Answer C**)

- First, identify the problem type: Proportional parts of a triangles.
- Thus, $\frac{BD}{DE} = \frac{BA}{AC}$ or $\frac{BA}{x}$
- $\frac{12}{9} = \frac{20}{AC} = \frac{20}{x}$
- 12x = 20 x 9 = 180
- x = 15
- Or $\frac{9}{x} = \frac{12}{20}$
- Cross multiply: 12x = 180; x = 15

Area of a Triangle

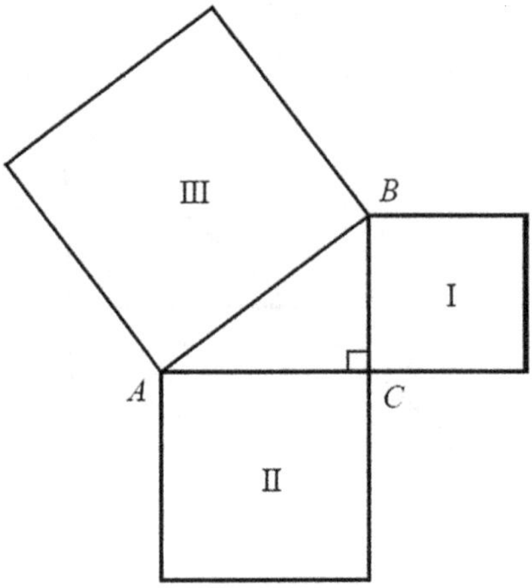

In the figure above, the area of square I is 85 and area of square II is 135? What is the area of square III?

A. Less than 220
B. Equals 220
C. More than 220
D. It cannot be determined from the information given

SAT Quick Solution (**Answer B**)

- Eliminate choice D. Mathematically choices A, B and C cover every possibility.
- Since Region III is a square, therefore, AB^2 is equal to the area of region III.
- We are given that the triangle is a right triangle. Thus, if we can find the length of AB then we can calculate the area of the square region III.
- If the area of region I is 85, then the length of side BC = $\sqrt{85}$
- If the area of region II is 135, then the length of side AC = $\sqrt{135}$
- By the Pythagorean Theorem: $A^2 + B^2 = C^2$
- $(\sqrt{85})^2 + (\sqrt{135}) = 85 + 135 = 220 = C^2$
- Thus, the area of region III is 220 because $C = \sqrt{220} = AB$
- The area of region III is $AB^2 = (\sqrt{220})^2 = 220$

Math Topic 18: Polygons

Polygons and Quadrilaterals

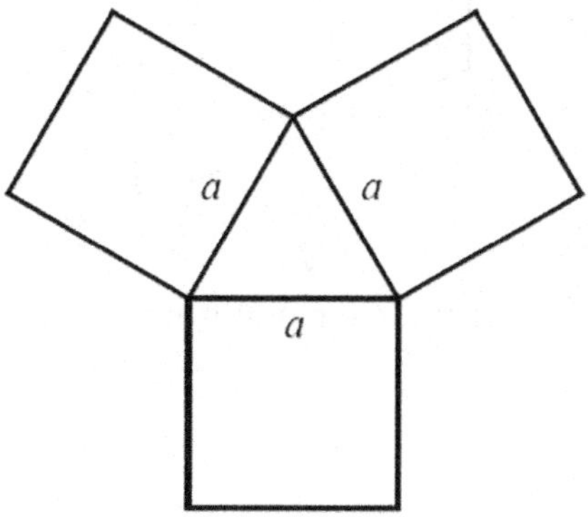

The equilateral triangle in the figure above has an area of $25\sqrt{3}$. What is the area of the 3 squares in the figure above?

A. 300

B. 360

C. 420

D. It cannot be determined from the information given

SAT Quick Solution (**Answer A**)

- Let's assume that you don't the formula to calculate the area of an equilateral triangle: $A = \frac{\sqrt{3}}{4}a^2$.
- Recommendation: **Commit Equilateral Triangle Formula to Memory**
- First, let's solve without using the formula for the equilateral triangle.
- We know the area of a triangle A = ½ bh
- Base b = a, and we can find the height using the Pythagorean Theorem
- $a^2 = (\frac{1}{2}a)^2 + h^2$, solving yields $h = \frac{\sqrt{3}}{2}a$. Do the math for yourself.
- Thus, we get the formula $A = \frac{\sqrt{3}}{4}a^2 = 25\sqrt{3}$ for this problem
- Solving yields $a^2 = 100$. We don't have to calculate **a** since we know $\mathbf{a^2}$ is the area of each square.
- Thus, the area of the 3 squares is 3 × 100 = 300.

Rectangles, Squares and Trapezoids

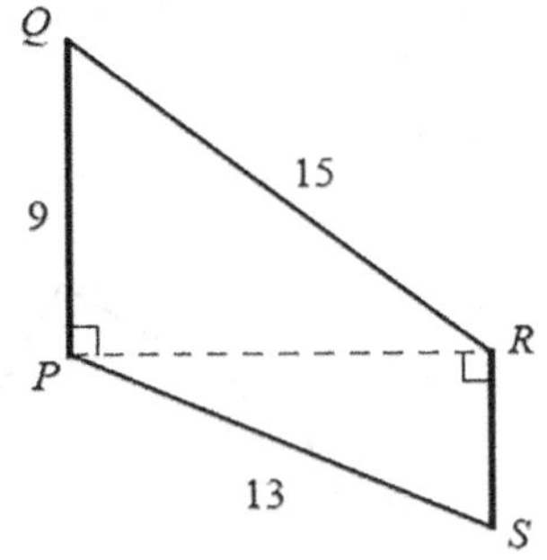

What is the area of the Trapezoid in the figure above?

A. 64
B. 72
C. 76
D. 84

SAT Quick Solution (**Answer A**)

- Find PR, 9, **12**, 15 is a Pythagorean Triple – learn those triples saves lots of time.
- **Need to Know**: Pythagorean Triples come in multiples. So, if you know 3, 4, 5 is a Pythagorean Triple, then you know that 9, 12, 15; 90, 120, 150; and 300, 400, 500 are Pythagorean Triples for example.
- **Advice**: Learn these Pythagorean Triples at a minimum:
    - **3, 4, 5;** 6, 8, 10; **5, 12, 13;** 8, 15, 17; **9, 12, 15;** 9, 40, 41
- Using another Pythagorean Triple (5, 12, 13), we know that RS = 5.
- Area of upper triangle = ½ bh = ½ (12) (9) = 54
- Area of lower triangle = ½ bh = ½ (12) (5) = 30
- Total area of the Trapezoid = 54 + 30 = 84
- SAT Approach
    - Note finding the height and RS as above no matter your approach.
    - The SAT Approach may be better if you remember the formula for area of a Trapezoid.
    - Use formula for Area of a Trapezoid = ½ $(b_1 + b_2) \cdot h$
    - Area of the Trapezoid = ½ $(5 + 9) \cdot 12 = 84$

Parallelogram

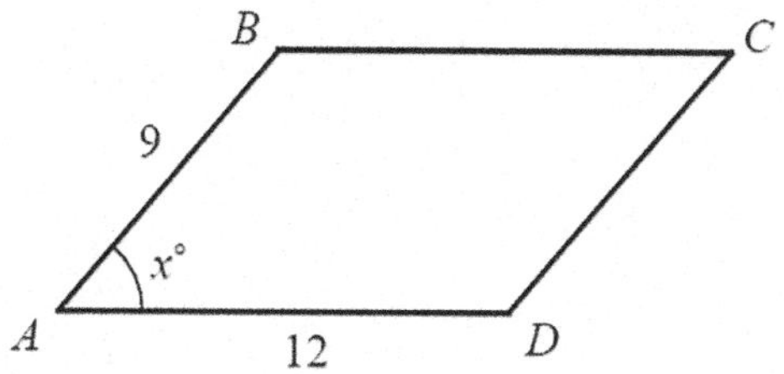

Below the figure, write the following: Find the Area of the Figure above.

A. 12 x 9 sin $x°$
B. $12 \times 9 \tan x°$
C. $12 \times 9 \sin x°$
D. $12 \times 9 \cos x°$

SAT Quick Solution (**Answer A**)

- **Need to Know:** The area of a parallelogram is bh
- The base b is 12
- We can calculate the height by drawing a vertical line from B to the base AD.
- 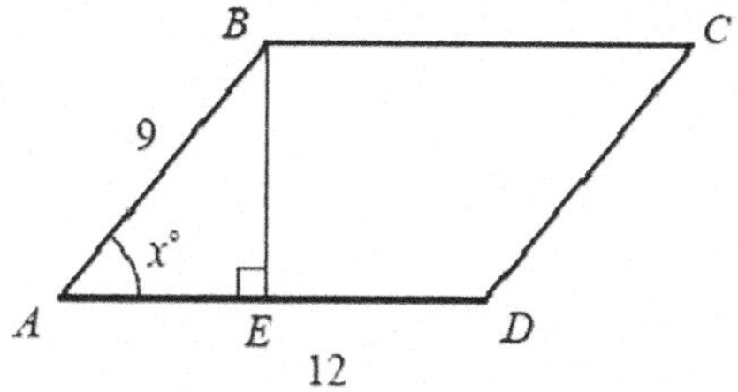

- Need to Know: $\sin = \frac{opposite}{hypotenuse}, \cos = \frac{adjacent}{hypotenuse}, \tan = \frac{opposite}{adjacent}$
- Apply sin because we need to find the height or opposite in this case
- $\sin x° = \frac{h}{9}$
- The height **h = 9 sin x°**
- Thus, **the area = 12 x 9 sin x°**

Math Topic 19: Circles

Arcs, Angles, and Tangents

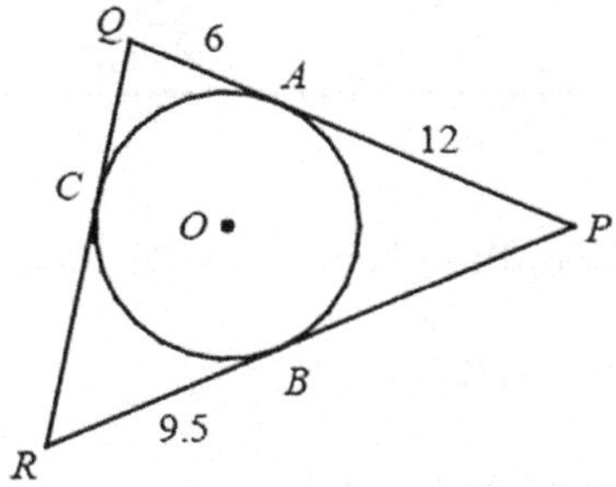

In the figure above, what is the perimeter of triangle PQR?

SAT Quick Solution (**Answer 55**)

- **Need to know: Tangents to a circle from the same exterior point are congruent**
- If you did not know this math fact, by inspection you could see that QA ≈ QC, AP ≈ BP, and CR ≈ BR. Sometimes common-sense works don't be frightened by the problem as Geometry & Trig problems are often straightforward.
- Therefore, in any case, QC = QA, AP = BP, and CR = BR
- Thus, the perimeter of Triangle PQR = 2(6 + 12 + 9.5) = 2(27.5) = 55

Arc Lengths and Areas of Sectors

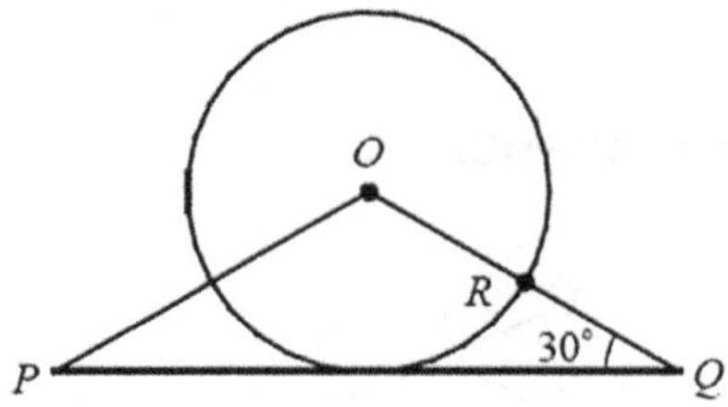

In the above figure, if the radius of circle O is 24, then what is the length of RQ?

SAT Quick Solution (**Answer 24**)

- **Apply rules for 30-60-90 Triangle**
- By the definition of a radius, the distance from O to the line PQ is 24.
- Therefore, by the rule for a 30-60-90 triangle, the length of OQ is 2 times the length of the shorter leg or OQ = 48
- Thus, to find RQ, we only need to subtract the radius: 48 − 24 = 24

Inscribed Angles

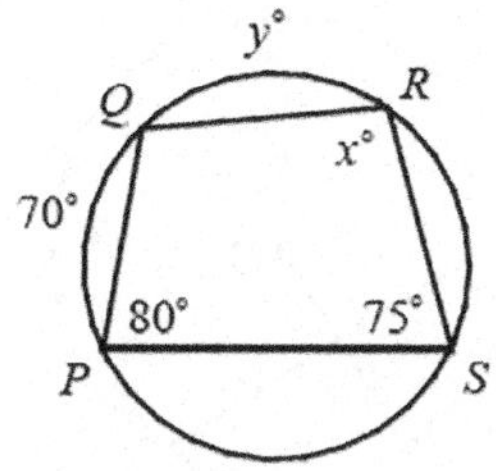

In the figure above, a quadrilateral is inscribed in a circle. What is the value of x?

SAT Quick Solution (**Answer 100**)

- Need to Know: **Opposite angles of a quadrilateral are supplementary**
- Thus, x + 80 = 180
- x = 100

Arcs and Chords

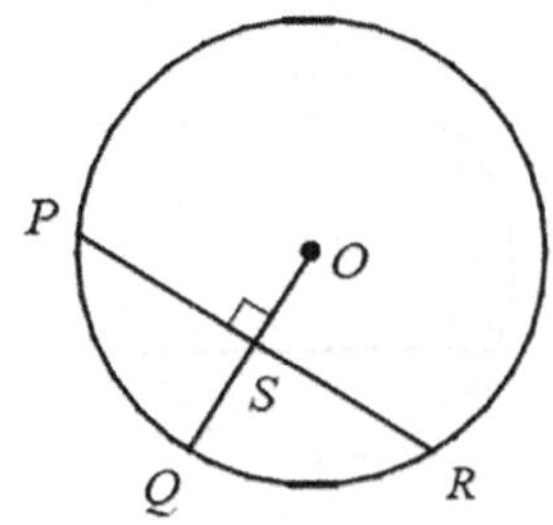

If the radius of the circle above is 13 and the length of PR is 24, then what is the length of QS?

SAT Quick Solution (**Answer 8**)

- OP = 13, PS = 12
- Apply the Pythagorean Triple **5**, 12, 13 to get the length of OS: 5
- QS + OS = 13
- QS + 5 = 13
- QS = 8
- SAT Standard Approach
    - OP = 13, PS = 12
    - $A^2 + B^2 = C^2$
    - $A^2 + 12^2 = 13^2$
    - $A^2 + 144 = 169$
    - $A^2 = 169 - 144$
    - $A^2 = 25$
    - A = 5 = OS, and QS + 5 = 13 such that QS = 8

Equations of Circles

$x^2 + 2x + y^2 - 4y - 9 = 0$

The equation of a circle in the plane is shown above. If the area of the circle is $k\pi$, then what is the value of k?

SAT Quick Solution (**Answer: k = 14**)

- Add 9 to both sides
- Take ½ of 2 = 1 and square it, then add it to 9 to get 10
- Take ½ or 4 = 2 and square it, then add it to 10 to get 14
- That's the value of r^2
- Thus, $k = r^2 = 14$
- SAT Standard Approach
    - Step 1: complete the square for the x-term: Add 1 to both sides
    - Step 2: complete the square for the y-term: Add 4 to both sides
    - $(x + 1)^2 + (y - 2)^2 + 1 + 4 - 9 = 1 + 4$
    - Add 9 to both sides
        - $(x + 1)^2 + (y - 2)^2 = 9 + 1 + 4$
        - $(x + 1)^2 + (y - 2)^2 = 14$
        - Realize the $r^2 = 14$
        - Area $= \pi r^2 = \pi (14) = 14\pi$
        - $k = 14$

Math Topic 20: Surface Areas and Volumes

Prisms

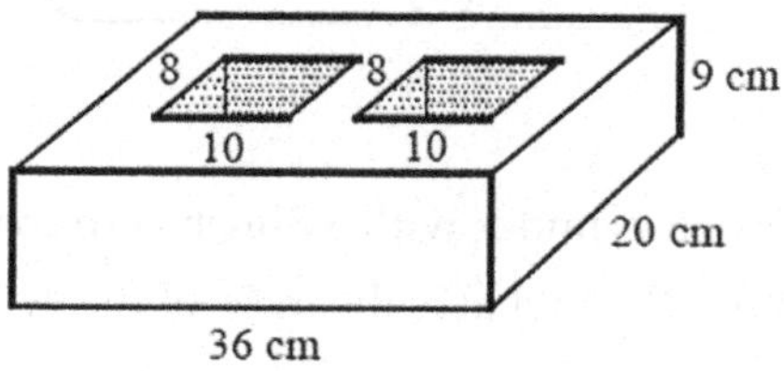

The cement block in the figure above is 36 cm × 20 cm × 9 cm and has two 10 cm × 8 cm holes. What is the weight of the cement block to the nearest gram, if the density of the cement is 1.7 gram / cm3?

SAT Quick Solution (**Answer: 8568 cm^3**)

- First calculate the entire volume as if there were no holes in the block.
  - $36 \times 20 \times 9 = 6480$ cm^3
- The volume of a single hole is $8 \times 10 \times 9 = 720$ cm^3
- 2 holes have a volume of $2 \times 720 = 1440$ cm^3
- The volume of the cement block is $6480 - 1440 = 5040$ cm^3
- Weight = density × volume = $5040 \times 1.7 = 8568$ cm^3

Cylinders and Spheres

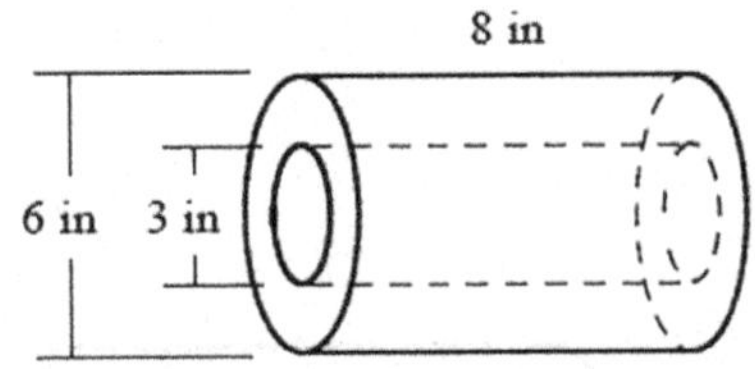

The figure above shows a cylinder with a 3-inch core carved out. If the density of the cylinder is 490 $\frac{lb}{ft^3}$, then what is the mass of the cylinder?

SAT Quick Solution (**Answer: 48 pounds**)

- Need to know: The volume of a cylinder, $V = \pi r^2 h$
- R = 3 inches for the entire cylinder
- r = 1.5 inches for the inner cylinder carved out
- Height, h = 8 inches
- Volume of the large cylinder, $V = \pi (3)^2(8) = 72\pi$
- Volume of the inner cylinder, $v = \pi (3/2)^2(8) = 18\pi$
- Thus, the volume of the figure above is $72\pi - 18\pi = 54\pi$ in^3
- Change to ft^3 → $54\pi \times \left(\frac{ft}{12\ inches}\right)^3 = 0.0982$
- Mass = density x volume = $490\frac{lb}{ft^3} x\ 0.0982\ ft^3 = 48\ pounds$

Pyramids and Cones

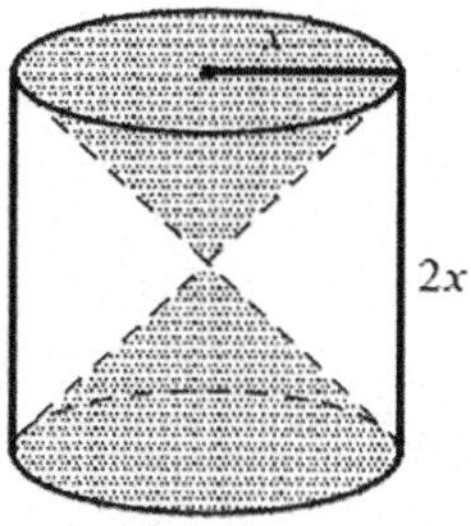

Two symmetrical cones are inscribed in the figure above. The radius of the cylinder is x and height of the cylinder is 2x. What is the volume of the cylinder if the two symmetrical cones are not included in the volume?

A. $\frac{1}{2}\pi x^3$
B. $\frac{2}{3}\pi x^3$
C. $\frac{4}{3}\pi x^3$
D. $\frac{3}{2}\pi x^3$

SAT Quick Solution (**Answer: C**)

- Use basic numbers: let x = 1 and 2x = 2
- Thus, the x^3 is 1 given that x = 1
- Volume of the cylinder, $V = \pi r^2 h = \pi(1)2(2) = 2\pi$
- Volume of a cone = $\frac{1}{3}\pi r^2 h = \frac{1}{3}\pi 1^2(1) = \frac{1}{3}\pi$
- Total volume of cones = 2 x volume of a single cone = $2(\frac{1}{3}\pi) = \frac{2}{3}\pi$
- Requested volume = $2\pi - \frac{2}{3}\pi = \frac{4}{3}\pi$
- Thus, the answer must be C.
- **SAT Standard Approach**
    - Volume of the cylinder, $V = \pi r^2 h = \pi(x)^2(2x) = 2\pi x^3$
    - Volume of a cone = $\frac{1}{3}\pi r^2 h = \frac{1}{3}\pi x^2(x) = \frac{1}{3}\pi x^3$
    - Total volume of cones = 2 x volume of a single cone = $2(\frac{1}{3}\pi x^3) = \frac{2}{3}\pi x^3$
    - Requested volume = $2\pi x^3 - \frac{2}{3}\pi x^3 = \frac{4}{3}\pi x^3$

SAT Math Test - No Calculator

1. The graph of y = $|x^3 - 1|$ is best represented by which of the following?

A. 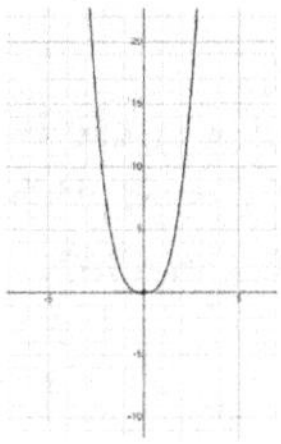

B. 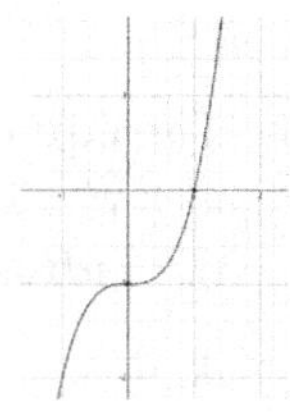

C. 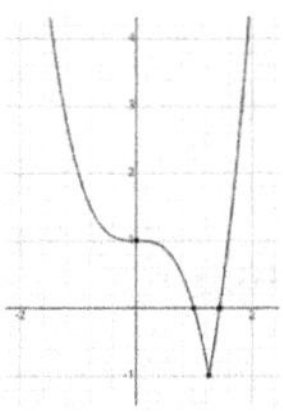

D. 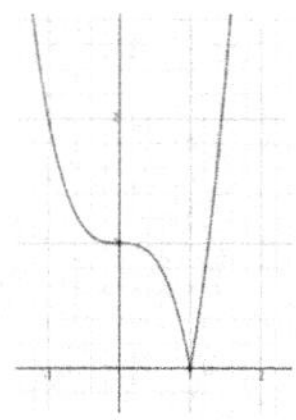

2. If 4x + 3y = 70 and 2x + y = 20, then what is 2x + 2y?

A. 25

B. 40

C. 50

D. 90

3. Which of the following statements is true?

A. $\frac{3}{8} < \frac{4}{11} < \frac{5}{13}$

B. $\frac{4}{11} < \frac{3}{8} < \frac{5}{13}$

C. $\frac{5}{13} < \frac{4}{11} < \frac{3}{8}$

D. $\frac{4}{11} < \frac{5}{13} < \frac{3}{8}$

4. Completely simplify

$$\frac{x^2-3x-4}{x^2+5x+6}\frac{x^2-4}{x^2-x-2}$$

A. 1

B. −1

C. $\frac{x+3}{x-4}$

D. $\frac{x-4}{x+3}$

5. In how many ways can the letters of the word PURSUE be rearranged such that the vowels always appear together?

A. 60

B. 72

C. 144

D. 360

6. If $m = n - 6$, then $(n - m)^2 + 12 = ?$

A. 36

B. 44

C. 48

D. 60

7. For the function f(x) = $ax^4 + 19$, if a is a constant and f (2) = 51, then what is f (−2)?

A. 0

B. −51

C. 51

D. 32

8. If 12 bananas cost b cents and 6 papayas cost p cents, then what is the cost of 4 bananas and 4 papayas in terms of b and p?

   A. $\frac{3b+p}{3}$

   B. $\frac{b+2p}{3}$

   C. $\frac{p+2b}{3}$

   D. $\frac{2bp}{3}$

9. What is a negative solution of $||x - 2| - 2| = 5$?

   A. 0

   B. −3

   C. −5

   D. −7

10. The average of several exam scores is 85. One make-up exam was given. Included with the other scores, the new average was 88. If the score on the make-up exam was 97, how many total exams were given?

    A. 2

    B. 3

    C. 4

    D. 5

11. If $x > 0$, how many integer values of will satisfy the equation $10x + 8|y| = 90$?

    A. 2

    B. 3

    C. 4

    D. 5

12. The equation of a circle is $x^2 + y^2 + 14x - 2y = -1$. What is the radius of this circle?

    A. 1

    B. 3

    C. 5

    D. 7

13. What is equal to $36^{\frac{3}{4}}$?

    A. $\sqrt[3]{36}$

    B. $\sqrt[4]{36}$

    C. 6

    D. $6\sqrt{6}$

14. A circle is inscribed in a square which is inscribed in a larger circle. What is the ratio of the area of the larger circle to smaller circle?

    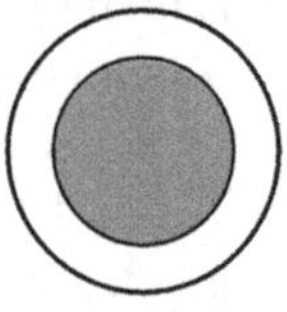

    A. $\sqrt{2}:1$

    B. 2:1

    C. $2\sqrt{2}:1$

    D. Cannot be determined

15. **Let A = the total area of 16 circles of radius r, and let B = the total area of 4 circles of radius a. If A = B, then what is ?**

A. 1

B. $\sqrt{2}$

C. 2

D. 4

16. **If o is an odd integer and e is an even integer, which of the following must be odd?**

A. $3e$

B. $o + 3$

C. $2(o + e)$

D. $o + 2e$

17. **What is the ones digit of 3^{719}?**

18. **When Sam paints a house, it takes him 5 hours. When Jenny joins Sam, and they work together, it takes them only 3 hours to paint a house of the same size. How long would it take for Jenny to paint a house of the same size by herself?**

19. **If $a : b : c = 3 : 4 : 7$ and $c - a = 12$, then what is $a + b + c$?**

20. **What is a positive root of $x^{1/2} - x^{1/4} - 2 = 0$?**

SAT Math Test - Calculator

1. The quadratic equation $x^2 + k^2x + 2k = 0$ has exactly one real solution for x. Given that k is a positive integer, what is the value of k?

   A. 1

   B. 2

   C. 3

   D. 4

2. The function $f(x) = x^2 - 8$ is defined for each of the following numbers EXCEPT

   A. 12

   B. 0

   C. −8

   D. −12

3. A radioactive substance has an initial mass of 10 grams, and its mass is reduced by 20% every 10 years. Which equation could be used to find the number of grams in the mass, y, that remains after x years?

   A. $y = 10(0.20)^{10x}$

   B. $y = 10(0.80)^{10x}$

   C. $y = 10(0.20)^{\frac{x}{10}}$

   D. $y = 10(0.80)^{\frac{x}{10}}$

4. Last month Becky worked 14 more hours than Harry. If they worked a combined total of 318 hours, how many hours did Becky work last month?

   A. 152

   B. 160

   C. 166

   D. 170

5. Line m in the xy-plane contains points from Quadrants I, II, III but no points from Quadrant IV. Which of the following must be true?

   A. The slope of line m is undefined

   B. The slope of line m is zero

   C. The slope of line m is positive

   D. The slope of line m is negative

6. What is $\frac{1}{19^{30}} - \frac{1}{19^{31}}$?

   A. $\frac{1}{19^{31}}$

   B. $\frac{30}{19^{31}}$

   C. $\frac{18}{19^{31}}$

   D. $\frac{9}{19^{31}}$

7. In a class of 78 students 45 are taking Statistics, 26 are taking Italian. Of the students taking Statistics or Italian, 11 are taking both courses. How many students are not enrolled in either course?

   A. 7

   B. 14

   C. 18

   D. 20

8. What is the area of the triangle below if the perimeter is 32?

   A. 40

   B. 48

   C. 56

   D. 64

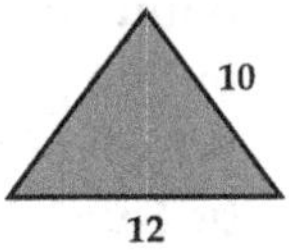

9. A system of three equations and their graphs in the *xy*-plane are on the left. How many solutions does the system have?

$$y = x^2 - 5$$
$$3y - 2x = 6$$
$$x^2 + y^2 = 25$$

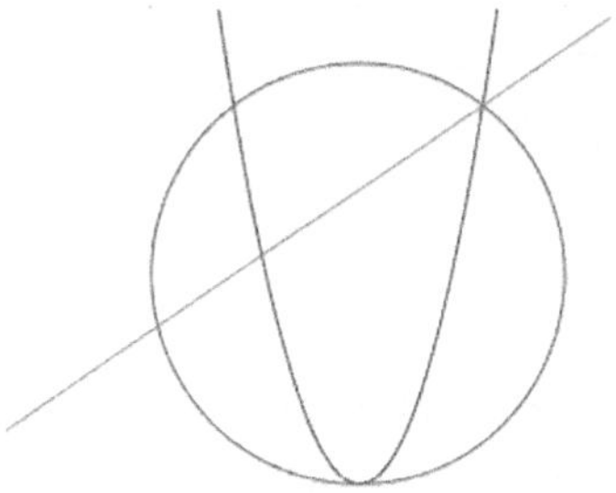

   A. 0

   B. 1

   C. 2

   D. 3

10. If the product of 10 integers is negative, what is the maximum numbers of integers that can be negative?

   A. 10

   B. 9

   C. 8

   D. 7

11. The perimeter of a triangle is 13. The lengths of the two shorter sides are m and m + 1. What is the length of the 3rd side?

A. 5

B. 6

C. 7

D. 9

12. A man runs m miles in s seconds. How many minutes will it take the man to run k miles?

A. $\frac{ks}{60m}$

B. $\frac{60km}{s}$

C. $\frac{60k}{ms}$

D. $\frac{k}{60ms}$

13. Syrup needs to be thinned to a ratio of 2 parts syrup to 1.5 parts water. The syrup maker has by mistake added water so that he has 6 gallons of syrup mixture which is half water and half syrup. What must he add to make the proportions of the mixture correct?

A. 1 gallon water

B. 1 gallon syrup

C. 1 gallon syrup and 1 gallon of water

D. ½ gallon syrup and ½ gallon water

14. The profits of ABC company rose 30% from January to April, then dropped 20% from April to September, then rose 50% from September through December. What was the percent increase for the whole year, from January through December?

A. 80%

B. 62%

C. 56%

D. 40%

15. Sam and Charlotte were looking to go on vacation to The Seychelles off the coast of Africa. The cost of the vacation was 60% more than Sam had in his vacation savings account, and separately, the same price was also 40% more than Charlotte had in her vacation savings account. What is the ratio of Sam's vacation savings account to Charlotte's savings vacation savings account?

A. 3:4

B. 5:6

C. 7:8

D. 8:7

16. You flip a fair coin 4 times. What's the probability that you will get a least one head?

A. $\frac{1}{2}$

B. $\frac{5}{8}$

C. $\frac{7}{8}$

D. $\frac{15}{16}$

17. If the measure of an exterior angle of a regular polygon is greater than 35 degrees, what is the greatest number of sides it can have?

A. 7

B. 8

C. 9

D. 10

18. If $x^{10}y^9 = 567$ and $x^8y^7 = 7$, then what is xy?

A. 3

B. 7

C. 9

D. 11

19. Mark forgot his bottle of water in the freezer, and it partially froze. He takes the water out of the freezer and sits it on his kitchen counter. The water changes temperature T in degrees Celsius according to the equation below where m represents time in minutes.

$T(m) = 20 - 22 \cdot 36^{-0.575m}$

What is the best interpretation of the number 20 for this scenario?

A. The bottle of water is originally 20 degrees Celsius.

B. Every 30 minutes, the temperature warms by 3 degrees Celsius.

C. After 20 minutes, the bottle of water will fully warm to the ambient temperature.

D. After the bottle of water has been sitting for a very long time, the temperature of the drink will approach 20 degrees Celsius.

20. If $15^6 + 15^7 = 2^x3^y5^z$, then what is the value of x + y + z?

A. 12

B. 14

C. 16

D. 18

21. The mean weight of the students in Fred's history class is equal to the mean weight of the students in his French class. However, while the standard deviation of the weights of students in his history class is 5.2 pounds, the standard deviation of the weights of students in his French class is 7.1 pounds. Which of the following statements must be true?

A. The mean weight of the students in Fred's French class is 1.9 pounds greater than the mean weight of the students in his history class.

B. The median weight of the students in Fred's French class is 1.9 pounds greater than the median weight of the students in his history class.

C. The median weight of the students in Fred's history class is 1.9 pounds greater than the median weight of the students in his French class.

D. The weights of the students in Fred's history class are less variable than the weights of the students in his French class

22. In a bag, there are four 6-sided dice (numbered 1 to 6), three 12-sided dice (numbered 1 to 12), and two 18-sided dice (numbered 1 to 18). If four of these dice are selected at random from the bag, and then the four are rolled and we find the sum of numbers showing on the four dice, how many different possible totals are there for this sum?

A. 57

B. 114

C. 570

D. 1140

23. Which of the following is a value of x for which the expression $\frac{x-3}{x^2-7x+10}$ is undefined?

A. 0

B. 5

C. −2

D. −3

24. If $v^{-\frac{1}{4}} = m$ where $v > 0$ what is in terms of m?

A. m^2

B. $-m^2$

C. $\frac{1}{m^4}$

D. $-\frac{1}{m^4}$

25. **When the integer n is divided by 6, the remainder is 3, Which of the following is NOT a multiple of 6?**

A. $n - 3$

B. $n + 3$

C. $2n$

D. $3n$

26. **If $4c = -\frac{1}{6}$, then what is $(1 - 120c)^2$.**

A. 24

B. 28

C. 32

D. 36

27. **In the figure given below, ABC and CDE are two identical semi-circles of radius 10 units. B and D are the midpoints of the arc ABC and CDE respectively. What is the area of the shaded region?**

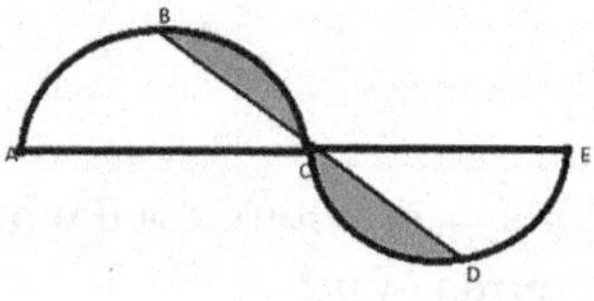

A. $100\pi - 25$

B. $75\pi - 25$

C. $50\pi - 100$

D. $25\pi - 100$

28. **What is the sum of the solutions of $x^2 - 3 = x$?**

A. 1

B. −1

C. $2(-1 + \sqrt{11})$

D. $2(1 - \sqrt{11})$

29. **If $n^3 = x$, $n^5 = 16x$ and $n > 0$, then what is the value of n?**

A. 1

B. 2

C. 4

D. 8

30. **Tom's average monthly cable bill for the months of January, February and March was $150. During which month did he receive the highest bill, if his lowest bill was received in February and the total of his bills for February and March was $320?**

A. January

B. February

C. March

D. Cannot be determined

31. For what value of k does the system of equations below have infinite solutions?

$$y = -\frac{k}{3}x + 4$$
$$3x + 9y = 36$$

32. 12 years ago, Aisha was half as old as Barack. Barack is now 10 years older than Aisha. How old will Barack be in 10 years?

33. If 2x + 1 = 2.5, then which of the following is 20x + 3?

34. What value of a will make the two lines below perpendicular to each other?

$$3x - 2y = 14$$
$$ax - 4y = 12$$

35. If $0 \leq x \leq 6$ and $-3 \leq y \leq 6$, what is the smallest possible value of $\frac{18}{|x-y|}$?

36. If the graph of $x^2 + y^5 = 25$ and the graph $y = b$ are drawn on the same coordinate plane, the graphs will not intersect, if b is less than what integer value?

37. Brenda takes the bus to visit her grandparents 300 miles from her home and returns with her grandfather. The average speed that her grandfather drives is 10 miles faster than the bus. If the total round-trip time is 11 hours, then how fast is Brenda's grandfather driving on average?

38. What is the length of a side, s, of a hexagon having an area equal to $96\sqrt{3}$?

SAT Math – No Calculator Section Solutions

1. **The graph of y = $|x^3 - 1|$ is best represented by which of the following?**

A. 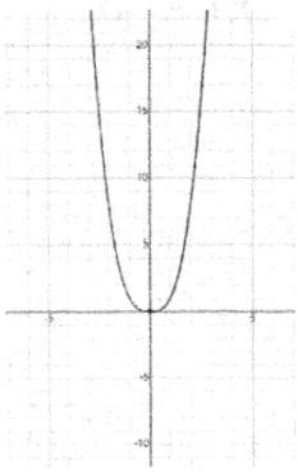

B. 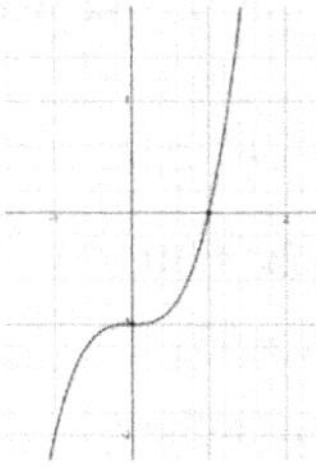

C. 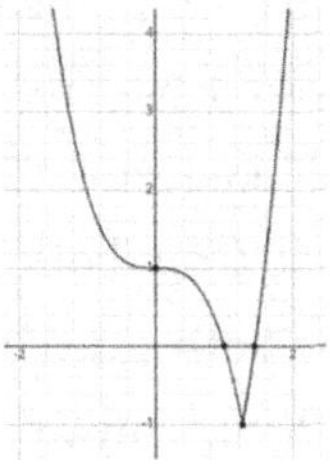

D. 

I. **Standard Solution (Answer Choice D)**

- First, find the y-intercept quickly: $y(0) = |0^3 - 1| = |-1| = 1$
- Only graphs C and D have a y-intercept (0,1).
- Thus, we can eliminate A, B.
- Secondly, the function is an absolute value function that can have no negative values which eliminates choice C.

II. **SAT Approach (Use Characteristics of the Different Graphs)**

- The absolute value function as given is always positive. Thus, we can eliminate Choices B & C.
- Graph A obviously represents a quadratic equation, which eliminates choice A

2. **If $4x + 3y = 70$ and $2x + y = 20$, then what is $2x + 2y$?**

I. **Standard Solution I (Answer Choice C)**

- Use substitution. Solve the second equation for y.
- $2x + y = 20$
- Subtract 2x from both sides yields: $y = 20 - 2x$
- Substitute for y in the first equation. $4x + 3(20 - 2x) = 70$
- Simplify: $4x + 60 - 6x = 70$
- Combine like terms: $-2x + 60 = 70$
- Subtract 60 from both sides: $-2x = 10$
- Divide by both sides by -2x: $x = -5$
- $y = 20 - 2(-5)$. Thus, $y = 20 + 10 = 30$
- $2x + 2y = 2(-5) + 2(30) = -10 + 60 = 50$

II. **Standard Solution II (Use Method of Elimination)**

- $4x + 3y = 70$ and $2x + y = 20$
- Multiply the second equation by -2: $-2(2x + y = 20) = -4x - 2y = -40$
- Add the 2 equations as seen in the textbook on the right.
- Now, solve the 2nd equation for x, $2x + 30 = 20$. Subtract 30 from both sides to get $2x = -10$ or $x = -5$
- Thus, $2x + 2y = 2(-5) + 2(30) = -10 + 60 = 50$

$$\begin{array}{r} 4x + 3y = 70 \\ -4x - 2y = -40 \\ \hline y = 30 \end{array}$$

III. **SAT Approach (Subtract the 2nd equation from the first equation)**

- When the SAT asks to evaluate an expression such as $2x + 2y$, it is frequently the case that you don't need to actually solve the equations in order to get the answer to problem. If you can spot such problems, it will save you considerable time.

$$\begin{array}{r} 4x + 3y = 70 \\ 2x + y = 20 \\ \hline 2x + 2y = 50 \end{array}$$

- If we subtract the second equation from the first equation (see above), we get $2x + 2y = 50$.

3. **Which of the following statements is true?**

A. $\frac{3}{8} < \frac{4}{11} < \frac{5}{13}$

B. $\frac{4}{11} < \frac{3}{8} < \frac{5}{13}$

C. $\frac{5}{13} < \frac{4}{11} < \frac{3}{8}$

D. $\frac{4}{11} < \frac{5}{13} < \frac{3}{8}$

I. **Standard Solution I (Answer Choice B)**

- **Find the common denominator**
- $8 \, x \, 11 \, x \, 13 \; = \; 1144$
- $\frac{3}{8}\left(\frac{11x13}{11x13}\right) = \frac{3x11x13}{8x11x13} = \frac{429}{1144}$
- $\frac{4}{11}\left(\frac{8x13}{8x13}\right) = \frac{4x8x13}{8x11x13} = \frac{416}{1144}$
- $\frac{5}{13}\left(\frac{8x11}{8x11}\right) = \frac{5x8x11}{8x11x13} = \frac{440}{1144}$
- **Timesaver**: Compare the numerators only since the common denominator is the same for 3 fractions.

II. **SAT Approach (Cross Multiply to Compare Fractions)**

- $3(11) > 4(8) \rightarrow 33 > 32, \, so \, \frac{3}{8} > \frac{4}{11}$
- Thus, $\frac{4}{11} < \frac{3}{8}$ *(Eliminates A)*
- Next examine Choice B: Is $\frac{3}{8} < \frac{5}{13}$?
- $3(13) < 5(8), \, so \frac{3}{8} < \frac{5}{13}$
- Thus, we can conclude that $\frac{4}{11} < \frac{3}{8} < \frac{5}{13}$
- **Tip**: Make sure that the answers relate to the numerator. In this example 33 relates to $3/8$ and 32 to $4/11$

4. Completely simplify $\frac{x^2-3x-4}{x^2+5x+6}\frac{x^2-4}{x^2-x-2}$

I. Standard Approach (Answer D)

- $\frac{x^2-3x-4}{x^2+5x+6}\frac{x^2-4}{x^2-x-2}$
- Factor all quadratic equations
- $\frac{(x-4)(x+1)}{(x+2)(x+3)}\frac{(x+2)(x-2)}{(x-2)(x-1)}$
- Cross out terms: $= \frac{(x-4)\cancel{(x+1)}}{\cancel{(x+2)}(x+3)}\frac{\cancel{(x+2)}\cancel{(x-2)}}{\cancel{(x-2)}\cancel{(x+1)}} = \frac{x-4}{x+3}$

II. SAT Approach (Substitute 0 for x)

- $\frac{x^2-3x-4}{x^2+5x+6}\frac{x^2-4}{x^2-x-2}$ (Choose 0 for x, it's quick and works if the denominator does not equal 0)
- If we let x = 0, we get the following:

  a. $-\frac{4}{6}\cdot\frac{-4}{-2} = -\frac{2}{3}\cdot 2 = -\frac{4}{3}$

- Since neither 1 nor -1 is equal to $-\frac{4}{3}$, we can eliminate answers A & B.
- If we substitute 0 for x in answer C, we get $-\frac{3}{4} \neq -\frac{4}{3}$. Only D is left!
- Of course, you should be able to quickly inspect and see that D is the correct answer when substituting 0 for x.

5. **In how many ways can the letters of the word PURSUE be rearranged such that the vowels always appear together?**

I. **Standard Solution (Answer B)**

- PURSUE = UUE PRS = UUE (treat UUE as a single unit)
- We have 4 entities in total = (UUE), P, R, S
- 4 entities can be arranged in 4 Factorial ways 4! = 4*3*2*1 = 24
- 3 vowels can be arranged in 3! ways if they were all different vowels.
- However, we have 2 U's. So, we must divide by 2! to account for the impact of the 2 U's.
- So total number of ways = $4! \times \frac{3!}{2!} = 24 \times 3 = 72$

II. **SAT Approach (Examine the problem as 4 different letters, then the Vowels next)**

- Since the vowels, must always appear together. Therefore, combine the vowels as a single unit.
- Let's call this unit V.
- **Find how many ways to arrange 4 letters**
- V P R S which can be arranged in four factorial ways: 4! = 4*3*2*1 = 24
- Find how many ways to arrange 3 vowels when one repeats.
- Since there are two U's and 1 E, they can be arranged UEU, EUU and UUE (3 ways)
- The total ways of arranging the alphabets of PURSUE = 21 x 3 = 72

6. **If $m = n - 6$, then $(n - m)^2 + 12 = ?$**

I. **Standard Approach (Answer C)**

- Given: $m = n - 6$
- Substitute for in the equation above yields and distribute the negative sign.
- $(n - m)^2 + 12 = (n - (n - 6))^2 + 12 = (n - n + 6)^2 + 12$
- $(6)^2 + 12 = 36 + 12 = 48$

II. **SAT Approach (Pick a number)**

- Given: $m = n - 6$
- Pick any number and evaluate the given equation to calculate m.
- Let $n = 8$, thus $m = n - 6 = 8 - 6 = 2$
- Now, let's substitute n and m to evaluate our expression
- $(8 - 2)^2 + 12 = 6^2 + 12 = 36 + 12 = 48$

1. **For the function f(x) = $ax^4 + 19$,**
if a is a constant and f (2) = 51, then what is f (−2)?

I. **Standard Approach (Answer C)**

- Solve for a and then find $f(-2)$
- f(2) $= a(2)^4 + 16 = 16a + 19 = 51$
- 16a + 19 = 51
- Subtract 19 from both sides 16a + 19 - 19 = 51 − 19 or
- 16a = 32
- $\frac{\mathbf{16a}}{\mathbf{16}} = \frac{\mathbf{32}}{\mathbf{16}} = 2$
- Thus, a = 2
- $\boldsymbol{f(-2) = 2(-2)^4 + 19} = 2 * 16 + 19 = \mathbf{32 + 19 = 51}$

II. **SAT Approach (Find $f(2)$ and $f(-2)$)**

- $f(2) = a(2)^4 + 19 = 16a + 19$
- $f(-2) = a(-2)^4 + 24 = 16a + 19$
- So, $f(2) = f(-2) \ = \ 16a + 19 \ = \ 51$
- No need to solve for a

III. **SAT Approach (Rely on the fact that x^4 is an even function)**

- *Since x^4 is an even function: $(x)^4 = (-x)^4 = x^4$*
- *Thus, changing x from 2 to −2 does not affect the solution since a is a constant (doesn't change)*
- $\therefore f(2) \ = \ f(-2) \ = \ 51$

8. **If 12 bananas cost b cents and 6 papayas cost p cents, then what is the cost of 4 bananas and 4 papayas in terms of b and p?**

I. **Standard Approach (Answer B)**

- If 12 bananas cost b cents, then 1 banana cost $\frac{b}{12}$ cents.
- If 6 papayas cost p cents, then 1 papaya cost $\frac{p}{6}$ cents.
- So, 4 bananas cost $4 \times \frac{b}{12} = \frac{4b}{12} = \frac{b}{3}$ cents.
- So, 4 papayas cost $4 \times \frac{p}{6} = \frac{4p}{6} = \frac{2p}{3}$ cents
- Total cost for 4 bananas and 4 papayas $= \frac{b}{3} + \frac{2p}{3} = \frac{b+2p}{3}$

II. **SAT Approach (Pick Costs and Use Answers)**

- First, always pick easy numbers that make the math simple.
- Assume the 12 bananas cost \$12.00 = b (*Use unrealistic cost of a banana to simplify math*)
- Assume the 6 papayas cost \$6.00 = p
- Based on the above assumptions 1 banana costs \$1.00 and 1 papaya costs \$1.00.
- So, 4 bananas and 4 papayas costs = 4(1.00) + 4(1.00) = \$8.00
- Next substitute for b and p in each of the answers until you get \$8 as an answer.
- $\frac{3b+p}{3} = \frac{3(12.00)+6.00}{3} = \frac{42.00}{3} = 14.00$
- $\frac{b+2p}{3} = \frac{12.00+2*6.00}{3} = \frac{24.00}{3} = 8.00$: **Eureka → Stop Here**

III. **SAT Approach (Use Proportions I)**

- If 12 bananas cost b cents, then 4 bananas cost $\frac{b}{3}$ cents
- If 6 papayas costs p cents, then 12 papayas cost 2p cents
- Thus, 4 papayas cost $\frac{2p}{3}$ cents.
- So, the total cost is $\frac{b}{3} + \frac{2p}{3} = \frac{b+2p}{3}$

IV. **SAT Approach (Use Proportions II)**

- If 12 bananas cost b cents, then 4 bananas cost $\frac{4}{12}b = \frac{1}{3}b = \frac{b}{3}$
- If 6 papayas cost p cents, then 4 papayas cost $\frac{4}{6}p = \frac{2p}{3}$
- So, the total cost is $\frac{b}{3} + \frac{2p}{3} = \frac{b+2p}{3}$

9. What is a negative solution of $|(|x-2|-2)| = 5$?

I. Standard Approach (Answer C)

- Eliminate A immediately, 0 is not a negative number!
- Solve $|(|x-2|-2)| = 5$
- $|x-2| - 2 = 5$
- $|x-2| = 7$
- $x - 2 = 7$
- $x = 9$
- *Always need to solve 2 equations for an absolute value problem.*
- $x - 2 = -7$
- *Add 2 to both sides*
- $x - 2 + 2 = -7 + 2$
- $x = -5$
- $x = -5$ is one of the answers given so we can stop here!!

II. SAT Approach (Use the Answers Given)

- *Eliminate A, 0 is not a negative number*
- Answers given indicated in red
- $||-3-2|-2| = ||-5|-2| = |5-2| = 3$
- $\mathbf{||-5-2|-2| = ||-7|-2| = |7-2| = 5}$
- −7: Not necessary to check −7 as an answer

III. SAT Approach (Use Cues from the Problem)

- What is a negative solution of $||x-2|-2| = 5$?
- Quick solution: You know that $7 - 2 = 5$. Also, you know that $|-7| = 7$. Therefore, if x is negative, then:
- $x - 2 = -7$
- Solving yields $x = -5$

10. The average of several exam scores is 85. One make-up exam was given. Included with the other scores, the new average was 88. If the score on the make-up exam was 97, how many total exams were given?

I. Standard Approach (Answer C)

- n = number of exams before make-up exam
- Total score when make-up exam is added = 85n + 97
- Total score after make-up exam is added also = 88(n + 1)
- 88(n + 1) = 85n + 97
- 88n + 88 = 85n + 97
- 3n = 9 or n = 3
- Total # of exams = n + 1 = 4

II. SAT Approach (Apply Assumption that the student makes 85 on all exams)

- Since the average is 85 to start. We can say that the student made an 85 on all 3 exams regardless of the student's actual score on the 3 exams.
- So, let make each 85 score and 88 score.
- 85 + 3 = 88
- 97 − 3 = 94 since I added 3 to an 85 score, I must subtract 3 from the 97 score. Repeat until all scores are 88!!!
- Let's do it again!!
- 85 + 3 = 88
- 94 − 3 = 91 since I added 3 to an 85 score, I must subtract 3 from the new 94 score
- One more time!!
- 85 + 3 = 88
- 91 − 3 = 88 since I added 3 to an 85 score, I must subtract 3 from the new 91 score
- Now, all scores are an 88.
- We raised to 3 exams to 88 and the lowered the 97 to 88.
- Thus, the total number of exams = 4

11. If $x > 0$, how many integer values of will satisfy the equation $10x + 8|y| = 90$?

I. **Standard Approach (Answer D)**

- You should be able to recognize when there are no integer solutions, but you can go through them systematically quickly if needed.
- **Let x = 1, yields 8 | y | = 80, 2 integer solutions: 10, -10**
- Let x = 2, then 8 does not go into 70, thus no integer solutions.
- Lex x = 3, then 8 does not go into 60, thus no integer solutions
- Let x = 4, then 8 does not go into 50, thus no integer solutions
- **Let x = 5, then 8 | y | = 40, 2 integer solutions: 5, -5**
- Let x = 6, then 8 does not go into 30, thus no integer solutions
- Let x = 7, then 8 does not go into 20, thus no integer solutions
- Let x = 8, then 8 does not go into 10, thus no integer solutions
- **Let x = 9, then 8 | y | = 0, 1 integer solution: 0**
- Thus, there are 5 integer solutions for x > 0.

II. **SAT Approach (Break down the characteristics of the problem)**

- $\mathbf{10x + 8|y| = 90}$
- First note that we are looking at integer solutions only!!
- **8 | y |** must be a multiple of **8** and **10x** must be a multiple of **10**
- So, what values of x leaves us a multiple of 8 for 8 | y | : **1, 5, 9**
- **For example, 10(1) + 8 | y | = 90, 8 | y | = 80, or y = ±10**
- However, for each value of x less than 9, 2 possible values of y exist because of the absolute value
- Therefore, there are 5 possible values for y

III. **SAT Approach (Rearrange the equation and then solve)**

- Rearrange the equation: $8|y| = 90 - 10x$
- We know that $8|y| > 0$
- Solution will exist when $90 - 10x$ is a multiple of 8.
- **90 − 10x is a multiple of 8y when x = 1, 5 or 9.**
- If x = 1, then 8 | y | = 80, thus y can be 10 or −10
- If x = 5, then 8 | y | = 40, thus y can be 5 or −5
- If x = 9, then 8 | y | = 0, thus y can only be 0.
- Therefore, there are 5 possible solutions.

12. **The equation of a circle is $x^2 + y^2 + 14x - 2y = -1$. What is the radius of this circle?**

I. **Standard Approach (Answer C)**

- The traditional approach to a problem of this type is to complete the square.
- Take half of the coefficient of x-term and square it and half of the coefficient of the y-term and square it. Remember that you must add these values to the right-hand side also.
- $x^2 + 14x + \left(\frac{14}{2}\right)^2 + y^2 - 2y + \left(-\frac{2}{2}\right)^2 = -1 + \left(\frac{14}{2}\right)^2 + \left(-\frac{2}{2}\right)^2$
- $x^2 + 14x^2 + 49 + y^2 - 2y + 1 = -1 + 49 + 1$
- $(x + 7)^2 + (y - 1)^2 = 49 = 7^2 = r^2, r = 7$

II. **SAT Approach (Focus on the Radius)**

- For this problem we only need to look at the leading coefficient of the x-term (14x) and the y-term (−2y), so 14 and −2, respectively.
- Divide by 2 and square and add to the constant on the right-hand side of the equation
- $-1 + \left(\frac{14}{2}\right)^2 + \left(-\frac{2}{2}\right)^2$
- $-1 + 7^2 + (-1)^2 = -1 + 49 + 1 = 49 = r^2$
- $r = \sqrt{49} = 7$

13. What is equal to $36^{\frac{3}{4}}$?

I. Standard Approach (Answer D)

- Manipulate exponents
- Remember $\sqrt{a \times a} = a$, $a^m \times a^n = a^{m+n}$ *and* $(x^m)^n = x^{mn}$
- $36^{\frac{3}{4}} = (6^2)^{\frac{3}{4}} = 6^{\frac{3}{2}} = \sqrt{6^3} = \sqrt{6 \times 6 \times 6} = \sqrt{6 \times 6}\sqrt{6} = 6\sqrt{6}$
- OR
- $36^{\frac{3}{4}} = (6^2)^{\frac{3}{4}} = 6^{\frac{3}{2}} = 6^1\, 6^{\frac{1}{2}} = 6 \times \sqrt{6} = 6\sqrt{6}$

II. SAT Approach (Expand Exponent and Simply)

- $36^{\frac{3}{4}} = \sqrt[4]{36^3} = \sqrt[4]{36 \times 36 \times 36} = \sqrt[4]{6 \times 6 \times 6 \times 6 \times 6 \times 6}$
- **Remember $\sqrt[4]{a \times a \times a \times a} = a$**
- $= 6\sqrt[4]{6 \times 6} = 6(36)^{\frac{1}{4}} = 6(6^2)^{\frac{1}{4}} = 6 \times 6^{\frac{1}{2}} = 6\sqrt{6}$

III. SAT Approach (Compare Exponents)

- We are looking at $36^{\frac{3}{4}}$?
- Apply $36^{\frac{1}{2}} = \sqrt{36} = 6$ and ¾ > ½. Thus, $36^{\frac{3}{4}} > 36^{\frac{1}{2}}$
- $\sqrt[3]{36} = 36^{\frac{1}{3}}$. $\frac{1}{3} \neq \frac{3}{4}$: ELIMINATE
- $\sqrt[4]{36} = 36^{\frac{1}{4}}$. $\frac{1}{4} \neq \frac{3}{4}$: ELIMINATE
- $6 = 36^{\frac{1}{2}}$. $\frac{1}{2} \neq \frac{3}{4}$: ELIMINATE
- $6\sqrt{6}$: Correct answer by process of elimination

14. A circle is inscribed in a square which is inscribed in a larger circle. What is the ratio of the area of the larger circle to smaller circle?

I. Standard Approach (Answer B)

- First, note that the length of diagonal of a square is $\sqrt{2}$s. Apply the principal for 45-45-90 triangles: the length of the hypotenuse is $\sqrt{2}$s
- Assume the radius of the small circle is r.
- Thus, the area of the small circle = πr^2
- The length of a side of the square is 2r (2 times the radius of the small circle).
- The length of the diagonal of the square is $2\sqrt{2}$r which is the diameter of the larger circle. So, the radius of the larger circle is $\frac{2\sqrt{2r}}{2}$ or $\sqrt{2}$r.
- Therefore, the area of the larger circle is $\pi(\sqrt{2}r)^2 = 2\pi r^2$
- So, the ratio is $\frac{2\pi r^2}{\pi r^2} = 2$. Thus, the ratio is 2:1.

II. SAT Approach (Pick Numbers)

- Let the radius of the small circle r = 10.
- Thus, the area of the small circle = $\pi(10)^2 = 100\pi$
- The length of the side of the square is 2(10) = 20.
- From that we can calculate the diagonal of the square which is twice the radius of the larger circle: $20\sqrt{2}$
- Thus, the radius of the larger is $10\sqrt{2}$
- So, the area = $(10\sqrt{2})^2\pi = 100 * 2\pi = 200\pi$
- Ratio = $200\pi : 100\pi \; or \; 2:1$

15. Let A = the total area of 16 circles of radius r, and let B = the total area of 4 circles of radius a. If A = B, then what is $\frac{a^2}{r^2}$?

I. Standard Approach (Answer D)

- Let A = the total area of 16 circles of radius r, and let B = the total area of 4 circles of radius a. If A = B, then what is $\frac{a^2}{r^2}$?
- Area A of 16 circles of radius r = $16 \times \pi r^2 = 16\pi r^2$
- Area of the 4 circles B = $4 \times \pi a^2 = 4\pi a^2$
- $\frac{Area\ of\ the\ 4\ circles}{Area\ of\ the\ 16\ circles} = \frac{4\pi a^2}{16\pi r^2} = 1$ => **Equals 1 because A = B**
- $\frac{a^2}{4r^2} = 1\ or \frac{a^2}{r^2} = 4$

II. SAT Approach (Apply Reasoning)

- We only need to compare the number of circles: $\frac{A}{B} = \frac{16}{4} = 4$

16. If o is an odd integer and e is an even integer, which of the following must be odd?

I. Standard Approach (Answer D)

- *o is odd and e is even*
- Multiplying an even number by any number always yields an even number. (*odd or even*) · *even* = *even*.
- 3 is odd. Adding two odd numbers always makes an even number. *odd* + *odd* = *even*
- Multiplying any number by 2 makes the number even. 2 (*o* + *e*) = *even*
- Answer! Adding an odd and an even number is always an odd number. o + 2e = o, since 2e is even

II. SAT Approach (Pick Numbers)

- Let o = 5 and e = 10
- A: 3e = 3(10) = 30 → even
- B: o + 3 = 5 + 3 = 8 → even
- C: 2(o + e) = 2(5 + 10) = 2(15) = 30 → even
- Of course, we don't need to verify D, but I include verification here for completeness
- D) o + 2e = 5 + 2(10) = 5 + 20 + 25 → odd

17. What is the ones digit of

I. Standard Approach (Answer 7)

- Calculator won't help you! Must be a pattern? Think like an SAT problem solver.
- $3^1 = 3$
- $3^2 = 9$
- $3^3 = 27$
- $3^4 = 81$
- $3^5 = 423$
- There's a pattern for the last digit: 3, 9, 7, 1, 3, 9, 7, 1, 3, …
- The pattern repeats in groups of 4.
- So, all we need to do is divide 719 by 4 and find the remainder
- $\frac{719}{4} = 179\ \textit{Remainder}\ 3$
- Thus, the ones digit of 3^{719} is 7
- **Actual Answer for informational purposes only:**
- *112248912839199169521104991484102990497550623020827858128468435946495494947126611479292115713235454039171003351631536736292871990969297055127349338581254058976246296106458145355346504780001590549753078045871088196236161933091567694585689929461077702662087958411159661415218110474826097548027342848339991854316133641936414670901359415519 33679467*

18. When Sam paints a house, it takes him 5 hours. When Jenny joins Sam, and they work together, it takes them only 3 hours to paint a house of the same size. How long would it take for Jenny to paint a house of the same size by herself?

I. Standard Approach (Answer 7.5 hours)

- Given: 3 hours to paint an entire with both working together
- Let y = The time it takes Jenny to paint a house alone
- $Sam's\ rate\ = \frac{1\ house}{5\ hours} = \frac{1}{5}\frac{house}{hour}$
- $Jenny's\ rate\ = \frac{1\ house}{y\ hours} = \frac{1}{y}\frac{house}{hour}$
- $\frac{1}{5} \circ 3 + \frac{1}{y} \circ 3 = 1 \rightarrow \frac{3}{5} + \frac{3}{y} = 1$
- Subtract $\frac{3}{5}$ from both sides.
- $\frac{3}{5} + \frac{3}{y} - \frac{3}{5} = 1 - \frac{3}{5}$
- $\frac{3}{y} = \frac{2}{5}$
- $y = \frac{15}{2} = 7.5$

II. SAT Approach (Apply Shortcut)

- *If you have the rate that it takes one person to complete a task and the total time to complete that two people can complete the task working together than you can use a simple short cut.*
- Multiply the two given times together: $3 \times 5 = 15$
- Subtract the two times: $5 - 3 = 2$.
- So, the time it take Jenny is $\frac{15}{2}$ = 7.5 hours

19. **If $a:b:c = 3:4:7$ and $c - a = 12$, then what is $a + b + c$?**

I. **Standard Approach (Answer 42)**

- Apply multiples
- Let a = 3x, b = 4x and c = 7x.
- We know that c − a = 12. Replace c with 7x and a with 3x to get
- $7x - 3x = 12$
- $4x = 12$
- $x = 3$
- a + b + c = 3x + 4x + 7x = 3(3) + 4(3) + 7(3) = 9 + 12 + 21 = 42

II. **SAT Approach (Systematic Trial and Error)**

- We are given $a : b : c = 3 : 4 : 7$
- The difference $c - a = 7 - 3 = 4 \neq 12$
- Let's double a, b, c which does not change the ratio.
- $6 : 8 : 14$. The difference $c - a = 14 - 6 = 8 \neq 12$
- Let's triple a, b, c, which again does not change the ratio.
- $9 : 12 : 21$. The difference $c - a = 21 - 9 = 12$ EUREKA!
- $Answer = 42 \rightarrow 9 + 12 + 21 = 42$

20. **What is the positive root of $x^{1/2} - x^{1/4} - 2 = 0$?**

I. **Standard Approach (Answer 16)**

- Note $x^{\frac{1}{2}} = \sqrt{x}$ *and* $x^{\frac{1}{4}} = \sqrt[4]{x}$
- *USE substitution: let* $u = x^{\frac{1}{4}}$, thus $u^2 = x^{\frac{1}{4}} \cdot x^{\frac{1}{4}} = x^{\frac{1}{2}}$
- Thus, our equation becomes → $u^2 - u - 2 = 0$. Factoring yields $(u - 2)(u + 1) = 0$
- Thus, u = 2 or u = −1. We are looking for a positive root, select u = 2, ignore u = −1.
- $u = 2 = x^{\frac{1}{4}}$. To solve for x, raise both side to the 4th power.
- $\left(x^{\frac{1}{4}}\right)^4 = x = 2^4 = 16$

II. **SAT Approach (Apply Math Reasoning)**

- **Hint**: Try numbers that we can easily calculate by hand? Remember it's the SAT, so the answer will not have a square root in it!!
- *Try* $x = 0$: $0^{\frac{1}{2}} - 0^{\frac{1}{4}} - 2 = 0 - 0 - 2 = -2 \neq 0$
- *Try* $x = 1$: $1^{\frac{1}{2}} - 1^{\frac{1}{4}} - 2 = 1 - 1 - 2 = -2 \neq 0$
- *Try* $x = 16$: $16^{\frac{1}{2}} - 16^{\frac{1}{4}} - 2 = 4 - 2 - 2 = 0$
- *We might have tried 16 because x must be a number that has a fourth root and a square root*

SAT Math – Calculator Section Solutions

1. **The quadratic equation $x^2 + k^2x + 2k = 0$ has exactly one real solution for x. Given that k is a positive integer, what is the value of k?**

   I. **Standard Approach (Answer B)**
   - The general form of a quadratic equation: $ax^2 + bx + c = 0$.
   - Before solving let's review a few basics
     - $\boldsymbol{b^2 - 4ac > 0}$ **: 2 solutions**
     - $\boldsymbol{b^2 - 4ac = 0}$ **: 1 solution**
     - $\boldsymbol{b^2 - 4ac < 0}$ **: No real solutions**
   - Another easily derived and a good fact to know for the SAT is as follows:
     - **The sum of the solutions of a quadratic equation is $\frac{-b}{a}$**
   - For exactly 1 solution, the discriminant must equal 0.
   - $b^2 - 4ac = k^4 - 4(1)(2k) = k^4 - 8k = 0$
   - $k^4 - 8k = k(k^3 - 8) = 0$
   - One solution is k = 0 but is invalid based on the constraints of the problem.
   - $k^3 - 8 = 0$
   - Adding 8 to both sides of the equation yields
   - $k^3 = 8$
   - Taking the cube root of both sides yields $\boldsymbol{k = 2}$
   - Check $(2)^3 = 8$

   II. **SAT Approach (Trial and Error using the Answers Given)**
   - Try k = 1, then the equation becomes $x^2 + x + 2 = 0$.
   - This is not factorable, so $k \neq 1$
   - Try k = 2, then the equation becomes $x^2 + 4x + 4 = 0$.
   - $\textit{This is factorable } x^2 + 4x + 4 = (x + 2)(x + 2) = 0$
   - Thus, k must be equal to 2

2. The function $f(x) = x^2 - 8$ is defined for each of the following numbers EXCEPT

I. Standard Approach (Answer D)

- Try $f(x) = x^2 - 8 = 12$; rearrangement yields $x^2 = 20$; solvable
- Try $f(x) = x^2 - 8 = 0$; rearrangement yields $x^2 = 8$; solvable
- Try $f(x) = x^2 - 8 = -8$; rearrangement yields $x^2 = 0$; solvable
- Therefore, D is the correct answer
- Proof: $f(x) = x^2 - 8 = -12$; rearrangement yields $x^2 = -4$, thus, no real solution

II. SAT Approach (Apply Mathematical Reasoning)

- Since $x^2 \geq 0$ for all values of x
- The minimum value of $x^2 - 8 = 0 - 8 = -8$
- $-12 < -8$, so, the correct answer is D

3. **A radioactive substance has an initial mass of 10 grams, and its mass is reduced by 20% every 10 years. Which equation could be used to find the number of grams in the mass, y, that remains after x years?**

I. **Standard Approach (Answer D)**

- The basic radioactive decay equation: $A = A_o$
- First figure out the decay rate: 1.0 – 0.20 = 0.80
- Thus, eliminate A & C immediately.
- Next, it decays 20% every 10 years. Thus, after 10 years the exponent should be 1. Thus, the correct exponent is $\frac{x}{10}$
- Answer D

II. **SAT Approach (Plug in numbers and Mathematical Reasoning)**

- If you don't understand these types of problems, plugging in numbers and little logic goes a long way.
- First, you know that you start with 10 grams and after 10 years you lose 20%, leaving with you 8 grams.
- Only answer D gives you 8 grams after 10 years:
- A: $y = 10(0.20)^{10x} = 10(0.20)^{10(10)} = \sim 0$
- B: $y = 10(0.80)^{10x} = 10(0.80)^{10(10)} = \sim 0$
- C: $y = 10(0.20)^{\frac{x}{10}} = 10(0.20)^{\frac{10}{10}} = 10(0.20)^1 = 2$
- **D:** $\mathbf{y = 10(0.80)^{\frac{x}{10}} = 10(0.20)^{\frac{10}{10}} = 10(0.80)^1 = 8}$

4. **Last month Becky worked 14 more hours than Harry. If they worked a combined total of 318 hours, how many hours did Becky work last month?**

   I. **Standard Approach (Answer C)**

   - Harry worked less hours than Becky.
   - Thus, let x = hours Harry worked
   - Then, x + 14 = hours Becky worked
   - So, x + x + 14 = 2x + 14 = 318
   - Solving yields x = 152
   - So, Becky worked 152 + 14 = 166.

   II. **SAT Approach (Apply Mathematical Reasoning)**

   - Since, there are only 2 people, we know that Becky must have worked more than half the 318 hours as well as 14 hours more than Harry
   - ½ × 318 = 159. **(Enables us to quickly to eliminate A)**
   - So, 160 can't be the right either **(Eliminates B since Becky works 14 hours more Harry)**
   - Try 166. If Becky worked 166, then Harry 152 **(318 – 156 = 152).**
   - 166 + 152 = 318

5. **Line *m* in the xy-plane contains points from Quadrants I, II, III but no points from Quadrant IV. Which of the following must be true?**

I. **Standard Approach (Answer C)**

- Draw a quick picture.

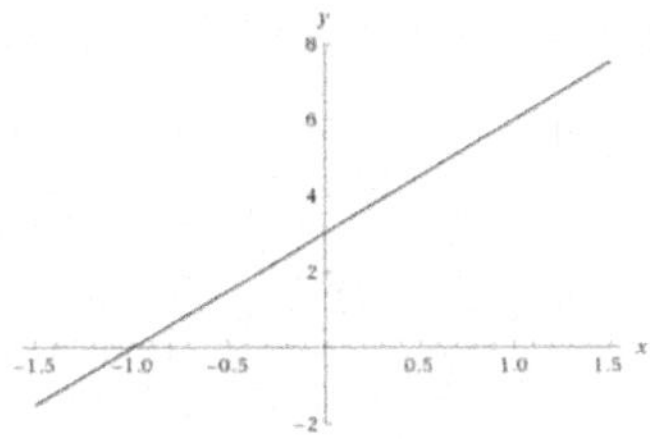

- The slope is positive

II. **SAT Approach (Visualize Graph Properties Mentally)**

- Eliminate A: An undefined slope is a vertical line and only exists in 2 quadrants.
- Eliminate B: A zero slope is a horizontal line which only exists in 2 quadrants.
- A positive slope rises from left to right and does not have to have any points in the 4th quadrant but can have points in quadrants I, II and III only.

6. **What is $\frac{1}{19^{30}} - \frac{1}{19^{31}}$?**

I. **Standard Approach (Answer C)**

- $\frac{1}{19^{30}} - \frac{1}{19^{31}}$
- **Make common denominator using the fact that 19 * 19^{30} = 19^{31}**
- $\frac{19}{19}\frac{1}{19^{30}} - \frac{1}{19^{31}} = \frac{19}{19^{31}} - \frac{1}{19^{31}} = \frac{19-1}{19^{31}} = \frac{18}{19^{31}}$

II. **SAT Approach (Compare Fractions)**

- First, notice that $\frac{1}{19^{31}}$ is much smaller than $\frac{1}{19^{30}}$, therefore the answer must be close to $\frac{1}{19^{30}}$
- Eliminate A since $\frac{1}{19^{31}}$ is not close to $\frac{1}{19^{30}}$
- Eliminate B since $\frac{30}{19^{31}} > \frac{1}{19^{30}}$
- $\frac{1}{19^{30}} = \frac{19}{19^{31}}$ which is close to $\frac{18}{19^{31}}$, therefore the answer is C

7. **In a class of 78 students 45 are taking Statistics, 26 are taking Italian. Of the students taking Statistics or Italian, 11 are taking both courses. How many students are not enrolled in either course?**

   I. **Standard Approach (Answer C)**

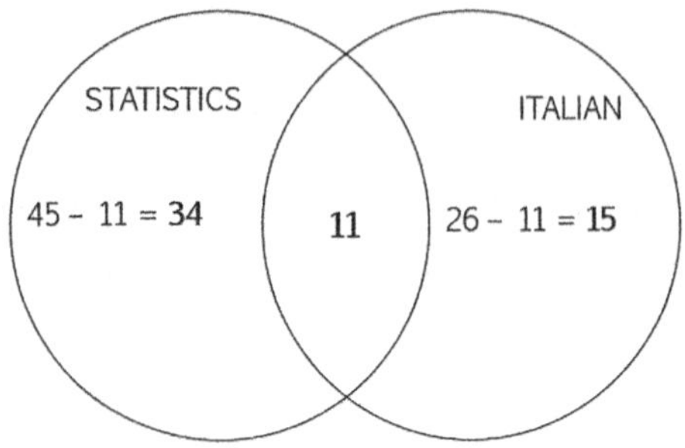

- Thus, those taking neither is 78 − 34 − 11 − 15 = **18**

II. **SAT Approach (Apply Mathematical Reasoning)**

- You can skip the vend diagram by following some simple logic.
- If there 78 students and 45 are taking statistics and 26 are taking Italian. Then, all you need to do is to subtract 45 and 25 from 78 and **add back in the 11 students taking both courses to cancel the double counting** for students taking both courses.
- 78 − 45 − 26 + 11 = 18

8. **What is the area of the triangle below if the perimeter is 32?**

I. **Standard Approach (Answer B)**

- Split the isosceles triangle into 2 right triangles

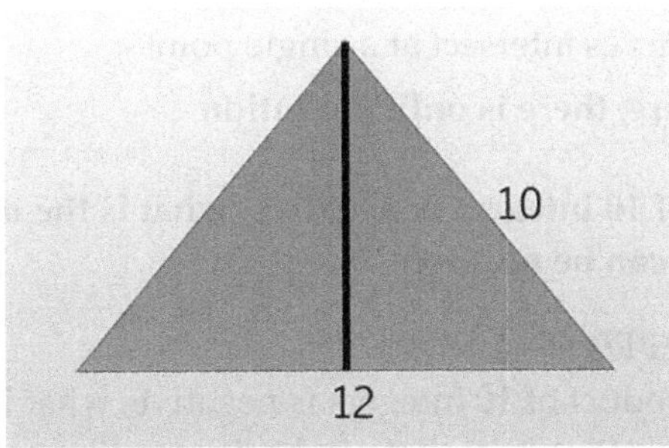

- Apply Pythagorean Theorem: $6^2 + h^2 = 10^2$
- $36 + h^2 = 100$
- $h^2 = 100 - 36 = 64$
- $h = 8$
- Area $= \frac{1}{2}bh = \frac{1}{2}(12)(8) = \mathbf{48}$

II. **SAT Approach (Apply Pythagorean Triples)**

- Split the isosceles triangle into 2 right triangles
- Learn your Pythagorean triples
- Common ones include: (3, 4, 5; 6, 8, 10; 5, 12, 13)
- The triangle is a 6-8-10 right triangle
- So, the height h = 8. Area $= \frac{1}{2}bh = \frac{1}{2}(12)(8) = \mathbf{48}$

III. **SAT Approach (Estimate)**

- Not the best approach, but it might it's worth a try.
- By inspection, the height must be less than 10.
- Assuming the height is 10, the area = ½ bh = ½ (12)(10) = 60
- Thus, the area is less than 60. Therefore, we can eliminate D.
- The height is probably an even number, so let's 8
- Using this approximation, the Area = ½ bh = ½ (12)(8) = 48

9. **A system of three equations and their graphs in the *xy*-plane are on the left. How many solutions does the system have?**

   I. **Standard Approach (Answer B)**

   - Understand curves
   - The 3 curves intersect at a single point
   - **Therefore, there is only 1 solution**

10. **If the product of 10 integers is negative, what is the maximum numbers of integers that can be negative?**

    I. **Standard Approach (Answer B)**

    - If the product of 10 integers is negative, what is the maximum numbers of integers that can be negative?
    - Remember the product of two negative numbers is positive
    - So, the answer is 9.
    - **−− −− −− −− −+**

    II. **SAT Approach (Apply Mathematical Reasoning)**

    - Since a negative number multiplied by a negative number is positive. Thus, we only need a single positive number.

    III. **SAT Approach (Use Numbers)**

    - Viable but not preferred approach.
    - Try a single positive number
    - $-1 \cdot -1 \cdot -1 \cdot -1 \cdot -1 \cdot -1 \cdot -1 \cdot -1 \cdot -1 \cdot 1 = -1$

11. The perimeter of a triangle is 13. The lengths of the two shorter sides are m and m + 1. What is the length of the 3rd side?

I. Standard Approach (Answer B)

- **Math Fact**: *For any triangle, the sum of the length of any 2 sides must be greater than the length of the 3rd side.*
- Define y = length of the 3rd side
- 1) $m + m + 1 = 2m + 1 > y$
- 2) $m + m + 1 + y = 2m + y + 1 = 13$
- 3) $2m + y = 12$
- 4) $2m = 12 - y$
- Why solve for 2m and not just m? The smart SAT taker doesn't waste time solving from m because 2m speeds up the solution process! 2m may be directly substituted into equation 1.
- Substitute Eq. 4 into Eq. 1 yields
- $13 - y > y$ **or** $y < 13 - y$
- $y < 13 - y$
- $+y \quad + y$
- $2y < 13 \quad or\ y < 6.5$
- The 3rd side has length 6 because 6 is largest integer less than 6.5.

II. SAT Approach (Trial and error)

- Use trial and error by taking advantage of the fact that the sum of the lengths of the 2 shorter sides is longer than the 3rd side.
- Let y = length of the 3rd side
- There's a much simpler way.
- Let m = 1, then m + 1 = 2,
- y = 13 − 1 − 2 = 10
- Not a triangle 1 + 2 < 13
- Let m = 2, then m + 1 = 3,
- y = 13 − 2 − 3 = 8
- Not a triangle $2 + 3 < 8$
- Let m = 3, then m + 1 = 4
- y = 13 − 4 − 3 = 6
- **A triangle $(3 + 4) = 7 > 6$**

12. **A man runs m miles in s seconds. How many minutes will it take the man to run k miles?**

I. **Standard Approach (Answer A)**

- Convert seconds to minutes multiply by $\frac{\textbf{1 minute}}{\textbf{60 seconds}}$
- Apply dimensional analysis. Quick an easy but some students may not understand this approach.
- $\frac{ks}{60m} = miles * seconds \frac{minutes}{seconds * miles} = minutes$

II. **SAT Approach (Pick Numbers)**

- **Hint**: Use numbers easily divisible by 60 since you are working with minutes and seconds.
- A man run m miles in s seconds. How many minutes will it take the man to run k miles?
- Use numbers. Let a man run 2 mile in 1200 seconds
- m = 2; s = 1200
- So, to run 10 miles, it would take 6000 seconds or 100 minutes, k = 10
- Plugging in these numbers yields A as the correct answer
- $\mathbf{10 * \frac{1200}{60*2} = 100}$ **minutes**

III. **SAT Approach (Apply Mathematical Reasoning)**

- A man run m miles in s seconds. How many minutes will it take the man to run k miles?
- To convert the seconds, s, to minutes we must divide s by 60.
- Thus, the answer must have $\frac{s}{60}$ in it.
- Therefore, we can eliminate B, C & D right away!!

13. Syrup needs to be thinned to a ratio of 2 parts syrup to 1.5 parts water. The syrup maker has by mistake added water so that he has 6 gallons of syrup mixture which is half water and half syrup. What must he add to make the proportions of the mixture correct?

I. Standard Approach (Answer B)

- Based on the problem statement the mixture contains:
- 3 gallons syrup and 3 gallons water
- However, the desired ratio is $\frac{2}{1.5}$ which means we need to add more syrup
- So, $\frac{3+s}{3} = \frac{2}{1.5}$ where s represents the syrup that must be added
- Cross-multiplying yields 1.5(3 + s) = 3(2)
- 4.5 + 1.5s = 6
- 1.5s = 1.5
- s = 1
- So, we need to add 1 more gallon of syrup!

II. SAT Approach (Apply Reasoning)

- Key: The 50/50 has too much water but the correct mixture has more syrup than water.
- Solution: Add more syrup only
- The only answer that fits this situation is Answer B

14. The profits of ABC company rose 30% from January to April, then dropped 20% from April to September, then rose 50% from September through December. What was the percent increase for the whole year, from January through December?

I. **Standard Approach (Answer C)**

- Let initial profits = p
- 30% increase = 1.3p
- 20% decrease = (1 − 0.2) * 1.3p = 1.04p
- 50% increase = 1.5 * 1.04p = 1.56p
- Over increases = $\frac{1.56p - p}{p} = 0.56 \text{ or } 56\%$

II. **SAT Approach (Apply Reasoning)**

- Let initial value of profits = p = 100
- 30% increase = 1.30 × 100 = 130
- *It's quicker to multiply by 1.30 to add 30% than to multiply by 0.30 and then add it back.* **Learn single step method to save time.**
- 20% decrease = (1.0 − 0.2) × 130 = 104
- 50% increase = 1.5 × 104 = 156
- The benefit of picking 100 enables you to calculate the 56% increase by inspection only but in case you cannot, here is the formula.
- Overall % increase = $\frac{156-100}{100} = \frac{56}{100} = 56\%$

15. Sam and Charlotte were looking to go on vacation to The Seychelles off the coast of Africa. The cost of the vacation was 60% more than Sam had in his vacation savings account, and separately, the same price was also 40% more than Charlotte had in her vacation savings account. What is the ratio of Sam's vacation savings account to Charlotte's savings vacation savings account?

I. Standard Approach (Answer C)

- Let s = Sam's savings and h = Charlotte's savings
- Let c = cost of vacation
- Sam: c = 1.6s
- Charlotte: c = 1.4h
- *Therefore,* $1.6s = 1.4h$ *or* $\frac{s}{h} = \frac{1.4}{1.6} = \frac{14}{16} = \frac{7}{8} = 7:8$

II. SAT Approach (Pick Numbers)

- Assume the vacation costs $224
- Why 16 x 14 = 224? It makes the math easier. However, if you don't recognize that this makes the math easier, then you may pick any number.
- So, Sam's savings is $\frac{224}{1.6} = 140$
- And, Charlotte's savings is $\frac{224}{1.4} = 160$
- $\frac{\text{Sam}}{\text{Charlotte}} = \frac{140}{160} = \frac{7}{8} = 7:8$

16. **You flip a fair coin 4 times. What's the probability that you will get a least one head?**

I. **Standard Approach (Answer C)**

- Use binomial probability. Long shot most students will not know or remember it. Also, to be frank, the SAT would not expect use of this approach.
- $P(x) = \frac{N!}{x!\,(N-x)!}\pi^x(1 - \pi)^{N-x}$
- N = number of trials
- x = number of successes
- $\pi = probability\ of\ success$
- In our case N = 4, $\pi = 0.5$
- The probability of flipping at least 1 head is 1 – probability of flipping no heads or $P(x \geq 1) = 1 - P(x = 0)$
- $P(x = 0) = \frac{4!}{0!\,(4!)}\ 0.5^0\ 0.5^4 = \frac{1}{16}$
- **Note, 0! = 1.**
- $\therefore P(x \geq 1) = 1 - P(x = 0) = 1 - \frac{1}{16} = \frac{15}{16}$

II. **SAT Approach (Write Possible Combinations)**

- In efficient, but viable approach.
- HHHH, HHHT, HHTH, HHTT, HTHH, HTTH, HTTT, THHH, THHT, THTH, THTT, TTHH, TTHT, TTTH, TTTT
- As shown on the left, flipping a coin 4 times gives you 16 possibilities.
- You get at least 1 head on every four flips except in the case that you get four tails.
- Thus, the probability of getting at least 1 head is $\frac{15}{16}$.

III. **SAT Approach (Simple Mathematical Reasoning)**

- $2^4 = 16$ possibilities when flipping a coin 4 times.
- Note that there is only one scenario in which you can get no heads: (TTTT)
- Thus, getting a head at least once must occur in the other 15 possibilities.
- Thus, the probability of getting at least 1 head is $\frac{15}{16}$.

17. **If the measure of an exterior angle of a regular polygon is greater than 35 degrees, what is the greatest number of sides it can have?**

I. **Standard Approach (Answer D)**

- The measure of the exterior angles of a regular polygon is always 360. Thus, nA = 360, where n is the number of sides and A is the measure of each exterior angle.
- $35n = 360.$
- $n = \frac{360}{35} = 10\frac{2}{7} \approx 10.3$
- **Since a polygon cannot have 10.3 sides, the answer is 10.**

II. **SAT Approach (Apply Reasoning)**

- The measure of the exterior angles of a regular polygon is always 360.
- You know that $\frac{360}{10} = 36$, so that a triangle with 10 sides has exterior measuring 36 degrees which is greater than 35.

18. **If $x^{10}y^9 = 567$ and $x^8y^7 = 7$, then what is xy?**

I. **Standard Approach (Answer C)**

- Breakdown $x^{10}y^9 = x^8x^2y^7y^2 = (x^8y^7)(x^2y^2) = 567$
- Substitute for x^8y^7: $7(x^2y^2) = 567$
- Thus, $x^2y^2 = (xy)^2 = \frac{567}{7} = 81$
- Thus, $xy = \sqrt{(xy)^2} = 9$

II. **SAT Approach (Apply Division)**

- Notice that $= \frac{x^{10}y^9}{x^8y^7} = x^{10-8}y^{9-7} = x^2y^2 = (xy)^2 = \frac{567}{7} = 81$
- Thus, $\sqrt{(xy)^2} = xy = 9$

19. Mark forgot his bottle of water in the freezer, and it partially froze. He takes the water out of the freezer and sits it on his kitchen counter. The water changes temperature T in degrees Celsius according to the equation below where m represents time in minutes.

$$T(m) = 20 - 22 \cdot 36^{-0.575m}$$

What is the best interpretation of the number 20 for this scenario?

I. Standard Approach (Answer D)

- $T(m) = 20 - 22 \cdot 36^{-0.575m}$
- Rewrite the equation: $T(m) = 20 - \frac{22}{36^{0.575m}}$
- For long times, $\frac{22}{36^{0.575m}}$ approaches 0. Why? The numerator grows extremely large as compared to the denominator.
- **Thus, after a long time, the water temperate will be 20 Celsius.**

II. SAT Approach (Apply Reasoning)

- Eliminate A. The bottle just came from the freezer so the original temperate is much less than 20 degrees Celsius.
- Eliminate B. If you set a frozen bottle of water on a counter, it is not going to warm at a rate of 3 degrees Celsius every 30 minutes
- Eliminate C. It's not realistic for a frozen bottle of water to thaw after 20 minutes.
- After a very long time, the bottle will reach the ambient/room temperature. Ice melts!!! Thus, the answer is 20 Celsius.

20. **If $15^6 + 15^7 = 2^x3^y5^z$, then what is the value of x + y + z?**

I. **Standard Approach (Answer C)**

- Learn those Exponent Rules: $\mathbf{(ab)^x = a^xb^x}$
- If $15^6 + 15^7 = 2^x3^y5^z$, then what is the value of x + y + z?
- **Simplify**
- $15^6 + 15^7 = (3{\cdot}5)^6 + (3{\cdot}5)^7 = 3^65^6 + 3^75^7 = 3^65^6 + 3^65^6{\cdot}3{\cdot}5$
- Factor the common term: $\mathbf{3^65^6}\cdot (1 + 3{\cdot}5) = 3^65^6 \cdot (16) = 2^43^65^6$
- x + y + z = 4 + 6 + 6 = 16

II. **SAT Approach (Use Common Factor)**

- Apply $x^n = x{\cdot}x^{n-1}$
- **Factor:** $\mathbf{15^6}(1 + 15) = 16 \cdot 15^6 = 2^4(3 * 5)^6 = 2^43^65^6$
- x + y + z = 4 + 6 + 6 = 16

21. **The mean weight of the students in Fred's history class is equal to the mean weight of the students in his French class. However, while the standard deviation of the weights of students in his history class is 5.2 pounds, the standard deviation of the weights of students in his French class is 7.1 pounds. Which of the following statements must be true?**

I. **Standard Approach (Answer D)**

- Students often have questions about standard deviation on the SAT but don't have experience with the mean of standard deviation.
- In simple terms, the standard deviation is a measure of variance. The smaller, the standard deviation, the closer the numbers are together.
- Therefore, the weights of the students in Fred's history class are less variable than the weights of the students in his French class.

22. **In a bag, there are four 6-sided dice (numbered 1 to 6), three 12-sided dice (numbered 1 to 12), and two 18-sided dice (numbered 1 to 18). If four of these dice are selected at random from the bag, and then the four are rolled and we find the sum of numbers showing on the four dice, how many different possible totals are there for this sum?**

    I. **Standard Approach (Answer A)**

    - The lowest possible total is always 4; all 1s.
    - The highest total would occur if we picked the two 18-sided dice and two of the 12-sided dice: 18 + 18 + 12 + 12 = 60.
    - Thus, we could have totals between 4 to 64, inclusive.
    - Consequently, we simply need inclusive counting:
    - 60 – 4 + 1 = 57

    II. **SAT Approach (Apply Mathematical Reasoning)**

    - You know the total is always greater than 0 and even if you pick the 4 highest dice, the total will never exceed 100. Consequently, there are less than 100 possible totals. Answer A.

23. **Which of the following is a value of x for which the expression $\frac{x-3}{x^2-7x+10}$ is undefined?**

    I. **Standard Approach (Answer B)**

    - An expression is undefined if the denominator equals 0. You cannot divide by 0.
    - Factor the denominator: $x^2 - 7x + 10 = 0$
    - $(x - 2)(x - 5) = 0$
    - Solutions: x = 2 or 5. The Answer is 5 (D) since 2 is not a choice.

    II. **SAT Approach (Apply Mathematical Reasoning)**

    - If x is negative, then the denominator is always positive, therefore eliminate choices C & D.
    - If x is 0, then the denominator equals 10, which eliminates choice C.
    - Thus, we are left B as the only possible answer.

24. If $v^{-\frac{1}{4}} = m$ where $v > 0$ what is v in terms of m?

I. **Standard Approach (Answer C)**

- Learn and apply exponent rules.
- Raise both sides to the $-\frac{1}{4}$ power:
- $\left(v^{-\frac{1}{4}}\right)^{-4} = v = m^{-4} = \frac{1}{m^4}$

II. **SAT Approach (Apply Mathematical Reasoning)**

- Remember v is greater than 0.
- Eliminate Choices B & D since they are always negative.
- There's a −¼ exponent in the problem, therefore both sides must be raised to the −4. Thus, the answer will not contain m^2

25. When the integer n is divided by 6, the remainder is 3, Which of the following is NOT a multiple of 6?

I. Standard Approach (Answer D)

- When n is split into groups of 6, 3 is left over.
- If you remove 3 from n, there will be only groups of 6 leftover. So (n − 3) will be divisible by 6.
- If you add 3 to n, the remaining will also form a complete group of 6 and hence there will be only groups of 6. So (n + 3) will be divisible by 6.
- When you double n, 3 left over will become 6 and so we will have only groups of 6. Hence 2n will be divisible by 6.
- When you triple n, 3 will still be left over. So 3n will not be divisible by 6

II. SAT Approach (Pick a number)

- **Try 9**: If 9 is divided by 6, then the remainder is 3.
- $9 - 3 = 6$ is a multiple of 6.
- $9 + 3 = 12$ is a multiple of 6.
- $2(9) = 18$ is a multiple of 6.
- $3(n) = 27$ is not a multiple of 6. Did not really need to check this one since the first 3 were multiples of 6.

26. If $4c = -\frac{1}{6}$, then what is $(1 - 120c)^2$.

I. Standard Approach (Answer D)

- If $4c = -\frac{1}{6}, then\ c = -\frac{1}{24}$
- $\left(1 - 120 * -\frac{1}{24}\right)^2 = (1 - (-5))^2 = (1 + 5)^5 = 6^2 = 36$
- When you triple n, 3 will still be left over. So 3n will not be divisible by 6

II. SAT Approach (Pick a number)

- $(30)4c = 120c = 30 * \left(-\frac{1}{6}\right) = -5$
- Common approach to transform equation not to solve equation first. In this case, 4c was transformed to 120c to make the calculation easier.
- $(1 - (-5))^2 = (1 + 5)^2 = 6^2 = 36$

27. In the figure given below, ABC and CDE are two identical semi-circles of radius 10 units. B and D are the midpoints of the arc ABC and CDE respectively. What is the area of the shaded region?

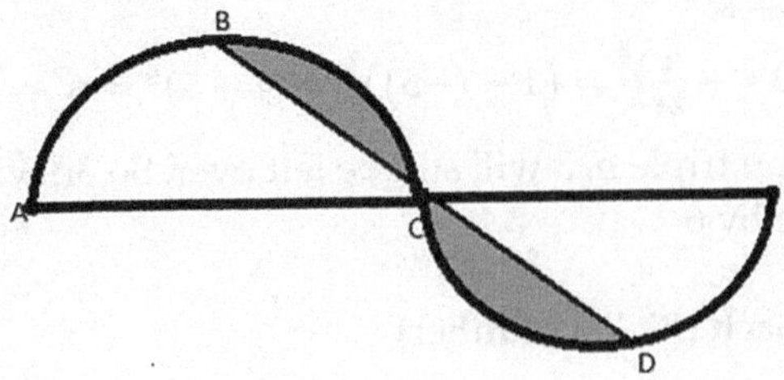

I. Standard Approach (Answer C)

- For problems such as this, break it up.
- There is a quarter of a circle, a semi-circle and a right triangle.
- Triangle Area = ½ (10)(10) = 50
- Quarter Circle Area = ¼ · Area of the circle = ¼ π $(10)^2$ = 25π
- Semi-circle Area = ½ π (10)2 = 50π
- Area of the shaded region = 50π − 25π − 50 = 25π − 50
- Now double it: 2(25π − 50) = 50π − 100

II. SAT Approach (Evaluate the different expressions)

- A: 100 π − 25 ≈ 286
- B: 75 π − 25 ≈ 210.5
- C: 50π − 100 ≈ 57
- D: 25 π − 100 ≈ -21.5
- The area of one semi-circle is ½ π $(10)^2$ = 50π ≈ 157
- The shaded area is obviously less than the area of the semi-circle which eliminate choices A & B.
- An area cannot be negative, which eliminates D.
- Thus, the only possible answer is C by method of elimination.

28. **What is the sum of the solutions of $x^2 - 3 = x$?**

I. **Standard Approach (Answer A)**

- What is the sum of the solutions of $x^2 - 3 = x$?
- Subtract x from both sides: $x^2 - 3 - x = x - x$
- Rearrange and simplify: $x^2 - x - 3 = 0$
- It's not factorable apply the quadratic equation:
- $a = 1, b = -1, c = -3$
- $x = \frac{-(-1) \pm \sqrt{(-1)^2 - 4(1)(-3)}}{2} = \frac{1 \pm \sqrt{13}}{2}$
- $\frac{1+\sqrt{13}}{2} + \frac{1-\sqrt{13}}{2} = \frac{1}{2} + \frac{1}{2} = 1$
- **The sum of two solutions always eliminates the square root term. Recognition of this math fact, automatically eliminates C & D.**

II. **SAT Approach (Understand Properties of the Quadratic Formula)**

- The quadratic equation: $x = \frac{-b \pm \sqrt{b^2 - 4ac}}{2a}$
- $a = 1, b = -1, c = -3$
- So, the sum of the solutions: $2\left(-\frac{b}{2a}\right) = -\frac{b}{a} = -\frac{-1}{1} = \mathbf{1}$

29. If $n^3 = x$, $n^5 = 16x$ and $n > 0$, then what is the value of n?

I. Standard Approach (Answer C)

- Substitute, Factor, then Solve
- If $n^3 = x,\ n^5 = 16x\ and\ n > 0$, then what is the value of n?
- Use substitution: $n^5 = 16x = 16n^3$
- Subtract $16n^3$ from both sides: $n^5 - 16n^3 = 0$
- Factor: $n^3(n^2 - 16) = 0$
- Ignore $n^3 = 0$, yields n = 0 as a solution
- $n^2 - 16 = 0 \rightarrow n = 4$

II. SAT Approach (Use Answers)

- If $\mathrm{n} = 1$, then $\mathrm{x} = 1$ but $1^5 = 1 \neq 16(1)$
- If $\mathrm{n} = 2$, then $\mathrm{x} = 8$ but $2^5 = 32 \neq 16(8)$
- **If $\mathbf{n} = 4$, then $\mathbf{x} = 64$** $\rightarrow 4^5 = 1024 = 16 * 64 = 1024$

30. Tom's average monthly cable bill for the months of January, February and March was \$150. During which month did he receive the highest bill, if his lowest bill was received in February and the total of his bills for February and March was \$320?

I. Standard Approach (Answer C)

- If the average for the 3 months is \$150, total bill for the 3 months is 3(150) = \$450
- If February and March total \$320, then January is \$130 by difference
- So, February must be lower than \$130, which means that March must be the highest: 320 − 130 = 190

31. For what value of k does the system of equations below have infinite solutions?

I. Standard Approach (Answer k = 1)

- For infinite solutions, the 2 equations must be the same **(parallel lines)**
- Step 1: Multiply the 1st Eq. by 9: **9y = -3kx + 36**
- Step 2: Rearrange: **3kx + 9y = 36**
- For the equations to be the same: **3k = 3 or k = 1**

II. SAT Approach (Use Common Slope Approach for Parallel Lines)

- For infinite solutions, the 2 equations must have the same slope.
- Slope of the 2nd Eq.: $\boldsymbol{9y = -3x + 36 \; or \; y = -\frac{x}{3} + 4}$
- *Slope of the 2nd Eq.:* $-\frac{1}{3}$; *Slope of the 1st Eq.:* $-\frac{k}{3}$
- $-\frac{1}{3} = -\frac{k}{3} \rightarrow k = 1$

32. 12 years ago, Aisha was half as old as Barack. Barack is now 10 years older than Aisha. How old will Barack be in 10 years?

I. Standard Approach (Answer 42)

- $a - 12 = \frac{1}{2}(b - 12) = \frac{b}{2} - 6$
- $b = a + 10$
- Multiply the first equation by 2: 2a − 24 = b − 12; 2a − 12 = b
- Thus: 2a − 12 = a + 10 −> a = 22 and b, Barack = 32.
- In 10 years, Barack will be 42.

II. SAT Approach (Guess numbers systematically)

- Try Barack = 24 now, then Aisha is 14 now.
- 12 years ago, Aisha was 2 and Barack was 12: 2 is not half of 12. Thus, Barack must be older.
- **24 is too young for Barack**
- Let Barack = 34 now, then Aisha is 24 now.
- 12 years ago, Aisha was 12 and Barack was 22: 12 is a little more than half of 22
- **Almost got it, let's try Barack = 32**
- Then Aisha is 22.
- 12 years ago, Aisha was 10 and Barack is 20.
- 10 is half of 20.
- **So, in 10 years, Barack will be 42.**
- **Honestly not an efficient approach but if solving equations is not your strong point it works.**

33. If $2x + 1 = 2.5$, then which of the following is $20x + 3$?

I. Standard Approach (Answer 18)

- $2x + 1 = 2.5$
- $2x = 1.5$
- $x = 0.75$
- $20(0.75) + 3 = 15 + 3 = 18$

II. SAT Approach (Guess numbers systematically)

- Transform the Equation, No Need to Solve
- $2x + 1 = 2.5$
- Multiple equation by 10: $20x + 10 = 25$
- Subtract 7 from both sides: $20x + 3 = 18$

34. What value of a will make the two lines below perpendicular to each other?

I. Standard Approach (Answer $\frac{8}{3}$)

- If two lines are perpendicular, then the product of their slopes is equal to −1
- What is the slope of the first line? **2y = 3x − 14 (m_1 = 3/2)**
- What is the slope of the second line? **4y = −ax + 12 (m_2 = −a/4)**
- $\frac{3}{2} \bullet \left(-\frac{a}{4}\right) = -\frac{3a}{8} = -1$ or -3a = -8
- Thus, $a = \frac{8}{3}$

II. SAT Approach (Apply Properties of the slope)

- **Trick**: If a line is in the form ax + by = c, then the slope, $m = -\frac{a}{b}$
- Slope of the first line: $m_1 = \frac{-3}{-2} = \frac{3}{2}$
- Slope of the second line: $m_2 = -\frac{a}{4}$
- For perpendicular lines, the product of the slope = −1
- $\frac{3}{2} \bullet \left(-\frac{a}{4}\right) = -\frac{3a}{8} = -1$
- $a = \frac{8}{3}$

35. **If $0 \le x \le 6$ and $-3 \le y \le 6$, what is the smallest possible value of $\frac{18}{|x-y|}$?**

I. **Standard Approach (Answer 2)**

- Evaluate $\frac{18}{|x-y|}$ at its endpoints:
- $|x-y| = |0-(-3)| = 3 \rightarrow \frac{18}{3} = 6$
- $|x-y| = |0-6| = 6 \rightarrow \frac{18}{6} = 3$
- $|x-y| = |6-(-3)| = 9 \rightarrow \frac{18}{9} = 2$
- $|x-y| = |6-6| = 0 \rightarrow \frac{18}{0}$ *is undefined*

II. **SAT Approach (Find largest possible denominator)**

- Realize $\frac{18}{|x-y|}$ is smallest when $|x-y|$ is largest:
- $|x-y| = |0-(-3)| = 3$
- $|x-y| = |0-6| = 6$
- $|x-y| = |6-(-3)| = 9$
- $|x-y| = |6-6| = 0$
- $\frac{18}{|x-y|} = \frac{18}{9} = 2$

36. If the graph of $x^2 + y^5 = 25$ and the graph $y = b$ are drawn on the same coordinate plane, the graphs will not intersect, if b is less than what integer value?

I. **Standard Approach (Answer −5)**

- For the graphs not to intersect $x^2 + y^2 \geq 25$
- Note that y = b is a horizontal line.
- We know that y = b, so now we $x^2 + b^2 \geq 25$
- $b^2 >= 25 - x^2$
- The maximum value of this inequality occurs when x = 0
- Therefore, $b^2 \geq 25$ or $b \geq \pm 5$
- −5 is the answer based on the question.

II. **SAT Approach (Understand the properties of the given circle)**

- This is a thought problem of sorts.
- First realize that the equation is a circle centered at the origin.
- The radius is 5, so the range is y = - 5 to y = 5.
- Thus, if b is less than -5, then the two equations will not intersect

37. Brenda takes the bus to visit her grandparents 300 miles from her home and returns with her grandfather. The average speed that her grandfather drives is 10 miles faster than the bus. If the total round-trip time is 11 hours, then how fast is Brenda's grandfather driving on average?

I. Standard Approach (Answer 60)

- If we let s = the speed of the bus, then Brenda's grandfather speed is s + 10.
- Remember time = t = $\frac{distance}{rate} = \frac{d}{r}$
- $\frac{300}{s} + \frac{300}{s+10} = 11$
- *Eliminate the denominator*
- **HOW**: *Multiply both sides of the equation by* $s(s+10)$
- $300(s+10) + 300s = 11s(s+10)$
- $300s + 3000 + 300s = 11s^2 + 110s$
- $600s + 3000 = 11s^2 + 110s$
- $11s^2 - 490s - 3000 = 0$
- Using quadratic equation yields
- s = 50 or −60/11 (Only need the positive solution)
- Remember to add 10 to the speed of the bus to get the grandfather's speed

II. SAT Approach (Use the estimated average speed as a guide point)

- The round trip is 300 + 300 = 600 miles
- It takes 10 hours in total, so the average speed = $\frac{600}{11} \sim 54.5\ mph$
- So, the speed of the bus must be less than 54.5 if that's the average speed for both
- Let's try 50 mph for the speed of the bus, so the grandfather travels at 60 mph
- Let's try 50 and 60 mph now
- $\frac{300}{50} + \frac{300}{60} = 11\ hours$
- So, Brenda's grandfather is driving 60 mph on average

38. What is the length of a side, s, of a hexagon having an area equal to $96\sqrt{3}$?

I. **Standard Approach (Answer 8)**

- Formula for the area of a hexagon: A = $3\frac{\sqrt{3}}{2}s^2$
- $A = 96\sqrt{3} = \frac{3\sqrt{3}}{2}s^2$
- $s^2 = 96\sqrt{3}\frac{2}{3\sqrt{3}} = \frac{96x2}{3} = 32x2 = 64$
- ∴ s = 8
- **The length of each side is 8 units**

II. **SAT Approach (Understand the properties of the given circle)**

- A hexagon is really 6 equilateral triangles as shown below

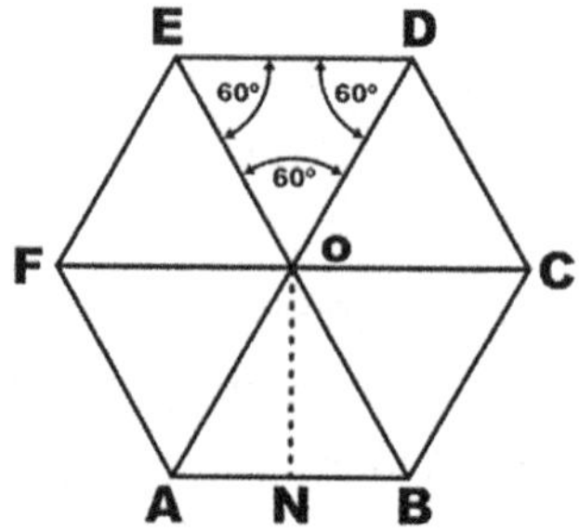

- We know the total area, divide that by 6 and get the area of a single triangle $\frac{96\sqrt{3}}{6} = 16\sqrt{3}$
- Use the formula for the area of an equilateral triangle: A = $\frac{\sqrt{3}}{4}s^2$
- A = $\frac{\sqrt{3}}{4}s^2 = 16\sqrt{3}$
- $s^2 = 16\sqrt{3} \times \frac{4}{\sqrt{3}} = 64$
- $s = 8$

III. Standard Approach (Apply Pythagorean Theorem)

- A hexagon is really 6 connected equilateral triangles
- Divide the total area by 6 to obtain the area of one triangle
  $\frac{96\sqrt{3}}{6} = 16\sqrt{3}$
- To find the height, split the triangle in half
- $h^2 + \left(\frac{s}{2}\right) = s^2; h^2 = s^2 - \frac{s^2}{4} = \frac{3}{4}s^2$
- $\therefore h = \frac{\sqrt{3}}{2}s$
- $A = \frac{1}{2}bh = \frac{1}{2}s\left(\frac{\sqrt{3}}{2}s\right) = \frac{\sqrt{3}}{4}s^2$
- *So,* $\frac{\sqrt{3}}{4}s^2 = 16\sqrt{3}$; *solving yields* $s = 8$
- **The length of each side is 8 units**

IV. SAT Approach (Use Trigonometry)

- For a 30-60-90 triangle, the long leg is the $\sqrt{3}$ *x the short leg*
- We know the total area, divide that by 6 and get the area of a single triangle $\frac{96\sqrt{3}}{6} = 16\sqrt{3}$.
- To find the height, split the triangle in half
- Use the rule for a 30-60-90 triangle
- The long leg (height) is $\frac{\sqrt{3}}{2}s$
- $A = bh = \frac{1}{2} \cdot s \cdot \frac{\sqrt{3}}{2}s = \frac{\sqrt{3}}{4}s^2$
- $\frac{\sqrt{3}}{4}s^2 = 16\sqrt{3}$
- $s^2 = 16\sqrt{3}\frac{4}{\sqrt{3}} = 64$
- s = 8
- The length of each side is 8 units